ASTRONOMY

Astronomy
ISBN 978-1-948211-87-1

Published by Quantum Scientific Publishing a division of Sentient Enterprises, Inc. Pittsburgh, PA. Copyright © 2017 by Sentient Enterprises, Inc. All rights reserved.

Permission in writing must be obtained from the publisher before any part of this work may be reproduced or transmitted in any form or by any means, electronic or mechanical, including photocopying and recording, or by any information storage or retrieval system. All trademarks, service marks, registered trademarks, and registered service marks are the property of their respective owners and are used herein for identification purposes only.

table of CONTENTS

Chapter 1 – The Solar System 1

Chapter 2 – The Formation of the Solar System 3

Chapter 3 – The Sun 5

Chapter 4 – Mercury 9

Chapter 5 – Venus 13

Chapter 6 – Earth 15

Chapter 7 – Mars 19

Chapter 8 – Jupiter 23

Chapter 9 – Saturn 27

Chapter 10 – Uranus 29

Chapter 11 – Neptune 31

Chapter 12 – Pluto 33

Chapter 13 – The Kuiper Belt 37

Chapter 14 – Comets 39

Chapter 15 – Past, Present, and Future Space Travel 41

Chapter 16 – The Galaxy 45

Chapter 17 – The Tools of Astronomers 49

Chapter 18 – The Hubble Space Telescope 53

Chapter 19 – The International Space Station 57

Chapter 20 – The Milky Way Galaxy 59

Chapter 21 – The Formation of the Milky Way Galaxy 61

Chapter 22 – Other Galaxies in the Universe 63

Chapter 23 – The Constellations 65

Chapter 24 – Stars 69

Chapter 25 – The Star Cycle 73

Chapter 26 – Supernovas 75

Chapter 27 – Quasars 77

Chapter 28 – Pulsars 79

Chapter 29 – Black Holes 81

Chapter 30 – Wormholes 83

Chapter 31 – The Universe 85

Chapter 32 – The Structure of the Universe 87

Chapter 33 – The Formation of the Universe 89

Chapter 34 – The Expanding Universe 91

Chapter 35 – The Big Bang Theory 93

Chapter 36 – Dark Matter 95

Chapter 37 – Background Radiation 97

Chapter 38 – The Doppler Effect 99

Chapter 39 – Galaxy Clusters 101

Chapter 40 – Voids in Space 103

Chapter 41 – Energy in the Universe 105

Chapter 42 – Einstein's Ideas 107

Chapter 43 – Special Relativity 109

Chapter 44 – Theories of Time Travel 111

Chapter 45 – Extrasolar Planets 115

Chapter 1 – The Solar System

Chapter Objective:

- Introduce the basic principles of astronomy, and the solar system's role in astronomy

Since the science of astronomy involves the study of physics, the interaction of the planets and their surrounding celestial bodies is directly related to the study of Astronomy. Astronomers look to the sky to be able to answer questions about our past and our future. These questions include understanding the stars and galaxies, the origins of the universe, the shape and structure of the solar system, time measurement, and so on. Study of these questions requires physics, including electromagnetic radiation, time, gravitational pull, and infrared radiation.

In early times, observations were made with the naked eye and interpretations and theories were developed based upon these physical observations. The early mapping of the stars and planets formed the basis for the early ideas of motion of the planets. In fact, it was believed that the Earth was the center of the universe and the Sun, planets, and stars rotated around it. This is called a geocentric view of the solar system and universe.

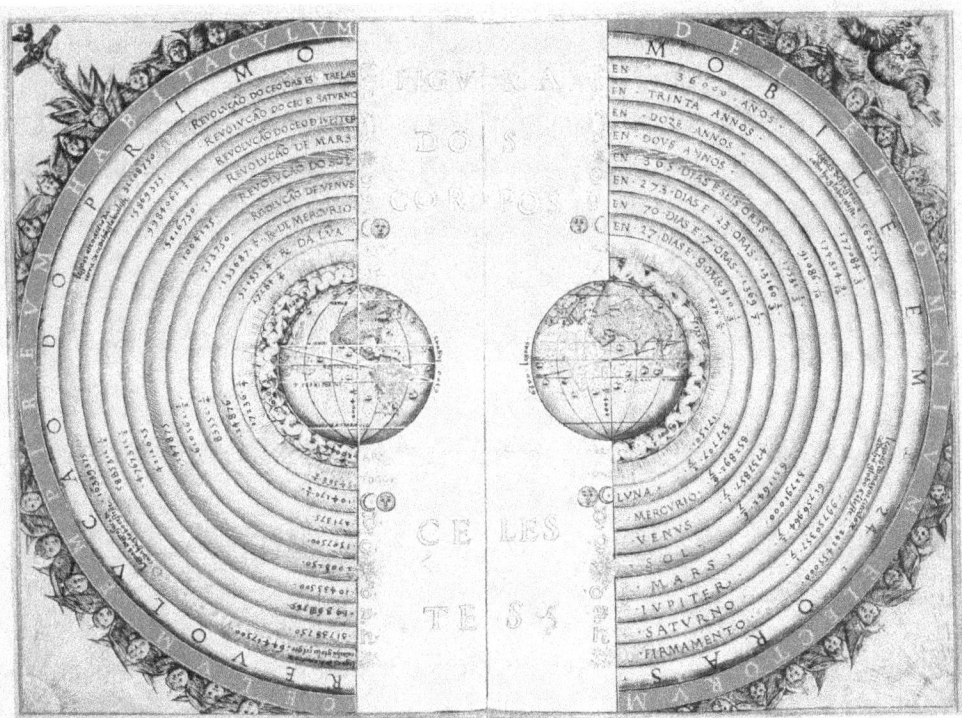

Figure of the heavenly bodies–Illuminated illustration of the Ptolemaic geocentric conception of the Universe by Portuguese cosmographer and cartographer Bartolomeu Velho (?-1568). From his work Cosmographia, made in France, 1568

Scientists have since learned that the planets actually orbit around the Sun in our heliocentric (sun centered) solar system.

The solar system is made up of a lot of empty space. The planets themselves are very small compared to the space in between them. So, how did the solar system form, millions of years ago?

Scientists believe the solar system was formed 4.6 billion years ago with the apparent gravitational collapse of a small part of a giant molecular cloud. The sun is the result of this collapsing mass collecting in the center. The remainder of the mass spread out into an elliptical disc, forming the planets, moons, asteroids, comets, and meteors that are present in the solar system.

Concept Reinforcement:

1. Describe questions about the universe that can be studied by astronomers using physics.

2. Discuss how astronomy has changed how people view the role of the earth in the solar system.

3. Explain the theory scientists have developed about the origins of the solar system.

Chapter 2 – The Formation of the Solar System

Chapter Objective:

- Apply the basic astronomy concepts to explain the formation of the solar system

Today the most accepted theory for how the solar system originated states that about 4.56 billion years ago the sun and planets formed from the solar nebula, which is a cloud of interstellar gas and dust.

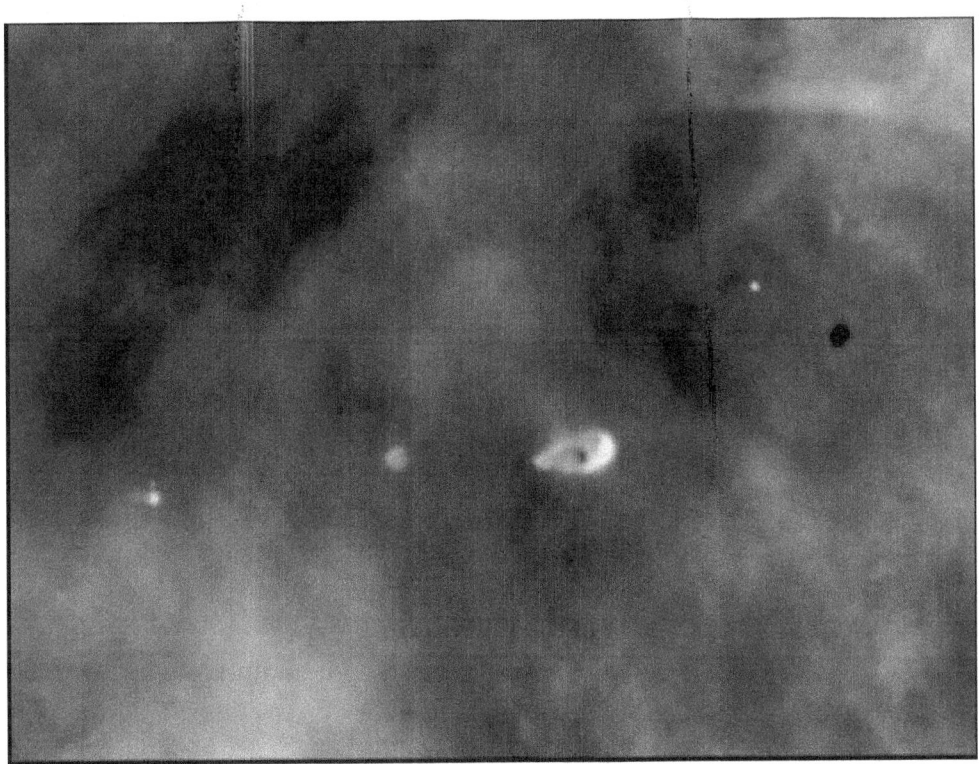

A Hubble Space Telescope view of a small portion of the Orion Nebula reveals five young stars. Four of the stars are surrounded by gas and dust trapped as the stars formed, but were left in orbit about the star. These are possibly protoplanetary disks, or "proplyds," that might evolve on to agglomerate planets. The proplyds which are closest to the hottest stars of the parent star cluster are seen as bright objects, while the object farthest from the hottest stars is seen as a dark object. Photo courtesy of NASA

Due to the fact that there are mutual gravitational attractions of material in the nebula, and this possibly was triggered by shock waves from a nearby supernova, the nebula collapsed on itself. As this nebula contracted, it spun ever more rapidly, leading to the development of a protostar, which was at the center of a protoplanetary disk. The dust and grains in the protoplanetary disk would orbit the protostar.

Artist's concept of a protoplanetary disk. Source: NASA

As the dust and grains collided frequently with each other, they combined together in a process of accretion, or bonding. These grains stuck together to form larger objects, first pebble sized, then boulder sized, until they formed into clumps of dust and grain that measured between 1 and 10 km in diameter. These clumps then collided to eventually form what are called planetestimals, which stuck to solid particles as well as gas. The planetestimals, according to the nebular hypothesis, then continued to collide with dust and grain, as well as larger bodies, over several million years. This led to the formation of protoplanets, which are the planets in their early stages. As the nebula continued to condense, the temperature at its core rose to the point where nuclear fusion began. This formation became the star we now call the sun, and the bodies father from the core became the planets of the solar system as we know it today. This theory is the modified version of the eighteenth-century nebular hypothesis.

Concept Reinforcement:

1. Describe the accepted theory of the formation of the solar system.

2. State the estimated age of the solar system.

3. Describe a solar nebula.

Chapter 3 – The Sun

Chapter Objective:

- Apply the basic astronomy concepts in relation to the sun, and the important role the sun has in the solar system

Our sun is an average sized star, which is classified as middle aged in relation to our galaxy, the Milky Way. The Sun is about 93 million miles away from the Earth. If we were to travel a bridge that went from the Earth to the Sun, it would take traveling at a constant 60 miles per hour for 177 years to go from one to the other.

The Sun is a gas ball made up primarily of hydrogen and helium, with a small amount of carbon, nitrogen, oxygen, and a mix of some heavy metals. The Sun is about 865,000 miles in diameter, which is about 109 times the diameter of the Earth. The Sun is large enough that over 1.3 million Earths could fit inside it and accounts for over 99 percent of the total mass of the solar system.

The Sun has several parts to its atmosphere. The first, outermost part is the Sun's corona, extending about 10 million miles out from the sun, and it is the outermost part of the sun's atmosphere. The corona is also the thinnest part of the Sun's atmosphere, and can only be viewed during a total eclipse of the Sun. Particles in the Sun's corona fade into a sea of charged particles that go into space as solar wind.

the Sun's Corona as seen during a Solar Eclipse

The Sun's chromosphere is the part of the atmosphere interior to the corona. This atmospheric layer of the Sun is accented by flares; bright and hot jets of gas, and faculae, which consists of bright hydrogen clouds known as plages.

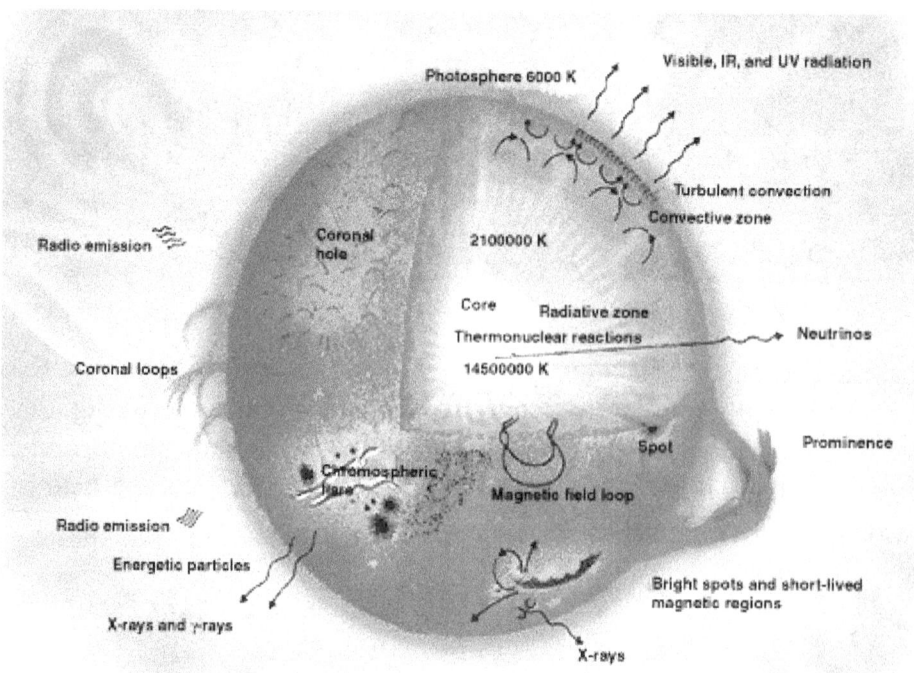
The Various Parts of the Sun. Image Courtesy of NASA

When we look at the Sun, what we actually see are the clouds of gases over its surface. This is a layer of the sun's atmosphere called the photosphere. The photosphere is a few hundred miles thick and is where the intense heat of the sun is given off into space. The photosphere is made up of Earth sized cells called granules. These cells are continually changing in size and shape as they carry hot gas from the center of the Sun to the surface, and cycle the cooler gas back down to the surface to be re-heated. It is within this layer that we see sun spots, or dark areas.

Below the Sun's atmosphere, interior to its photosphere, is the convection zone. In this region of the Sun, heat is carried towards the corona by slow moving gas currents. After the convection zone is the Sun's radiative zone, the region where heat is dispersed into the surrounding hot plasma. The core of the Sun is about 312,000 miles below its surface. It accounts for only 3 percent of the Sun's volume, but it is very dense, accounting for about 60 percent of the Sun's mass.

The temperature of the Sun's core is about 27 million degrees Fahrenheit (15 million °C). This is where the Sun's heat producing process, nuclear fusion, takes place. Fusion fuses four hydrogen nuclei into one helium nucleus. The fusion process releases energy.

The Sun will still be around for another 5 billion years, but it will spend roughly the final 10 percent of its life as a red giant. The surface temperature of the Sun will drop to between 3,000 and 6,700 degrees Fahrenheit (between 1,649 and 3,704 °C), and take on a reddish color. The diameter of the Sun will expand ten times its current size, engulfing the Earth in the process. The Sun's atmosphere will then begin to drift away after about a billion years, and leave a glowing core called a white dwarf. It will most likely take another trillion years to cool.

Concept Reinforcement:

1. Describe the composition of the Sun's atmosphere.

2. Explain the zones of the body of the Sun.

3. Discuss the process that the Sun uses to generate heat.

Chapter 4 – Mercury

Chapter Objective:

- Apply the basic astronomy concepts in relation to the planet Mercury, and the important role Mercury has in the solar system

Mercury is a small planet, located the closest to our Sun. Mercury is the smallest planet in our solar system. Pluto once held that distinction, but is now classified as a dwarf planet. The diameter of Mercury is over one third that of Earth's, but Mercury has only a little over 5 percent the mass of Earth. Mercury is 36 million miles from the Sun. The Sun exerts a strong gravitational pull on Mercury, forcing the orbit of Mercury to tilt into a long ellipse.

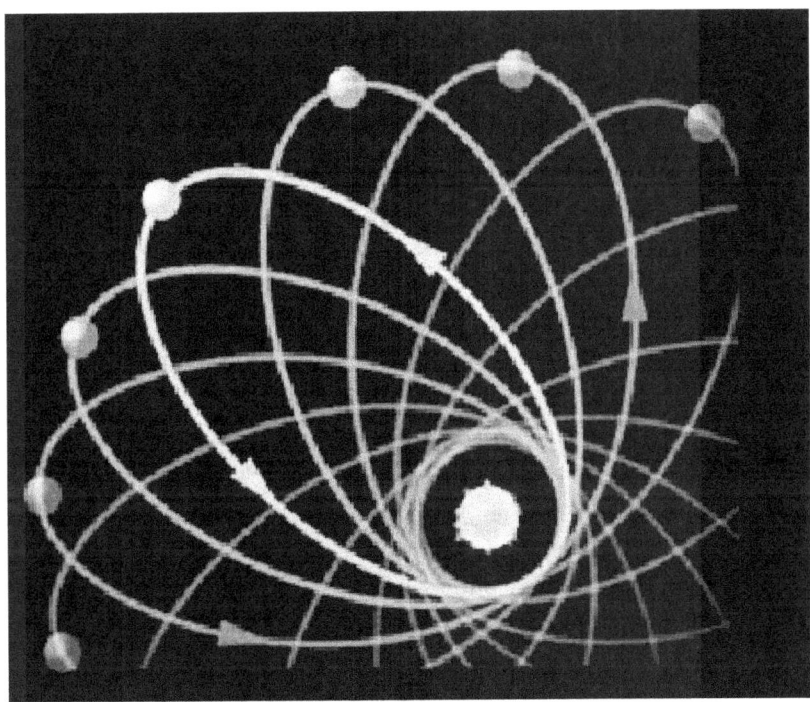

The Orbit of Mercury Around the Sun

Mercury was named for the Roman messenger god with winged sandals. Mercury orbits the sun very quickly, in just 88 days. However, Mercury has a very long day compared to the day we experience on Earth, which is equivalent to 59 Earth days for one rotation of the planet.

Astronomers believe that Mercury was originally made of liquid rock, like the moon. As the rock solidified, the planet cooled. Some meteorites hit the surface of the planet during the cooling, which formed craters.

The plains on the surface were formed by meteorites that broke through the cooling crust, causing lava to flow up to the surface, covering the older craters.

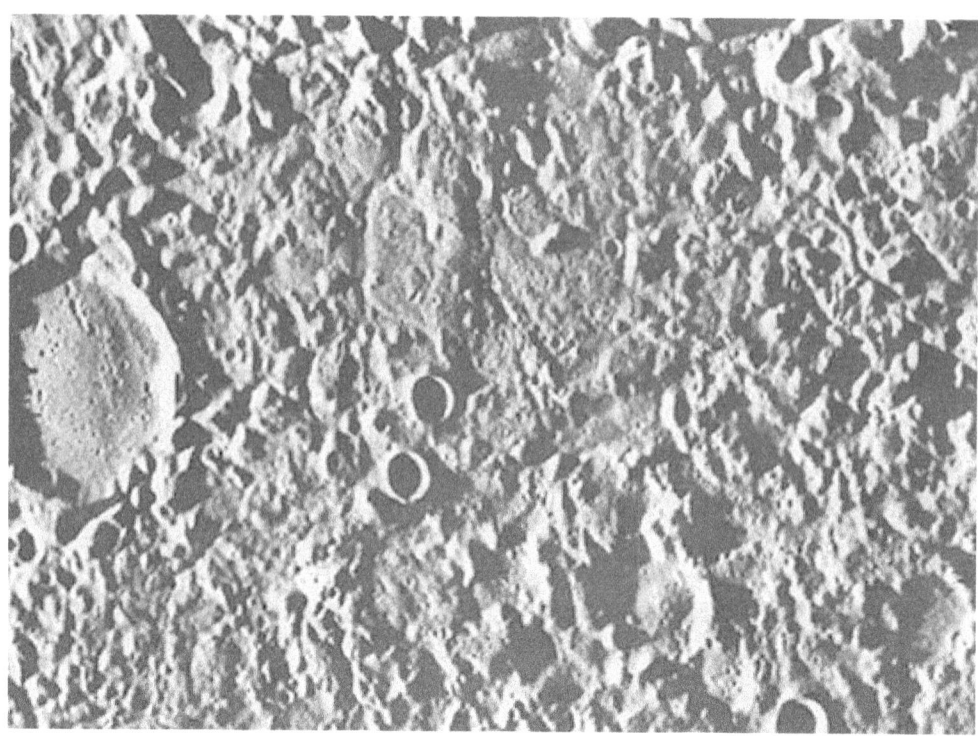
Mercury's Terrain image courtesy of NASA

It is hard to see Mercury with the naked eye due to the intense glare from the Sun. Mercury is usually only visible just above the horizon for one hour before sunrise or after sunset.

It wasn't until the space probe Mariner photographed Mercury, in 1975, that we knew anything about this planet. The probe took photos that showed that the surface of the planet is covered with craters, separated by plains and cliffs.

The mariner probe was also able to gather information about the composition of Mercury's core, which consists primarily of iron and nickel. Mercury's core is the densest of any of the planets in our solar system. The composition of the core is what probably protects the planet from the sun's particle wind.

The atmosphere of Mercury is very thin and made up of sodium, potassium, helium, and hydrogen. Temperatures on Mercury reach 800 degrees Fahrenheit (426 °C) during its long day and -280 Fahrenheit (-173 °C) during its long night. There are currently no space missions planned for Mercury.

The Launch of Mariner 1 image courtesy of NASA

Concept Reinforcement:

1. Describe Mercury's position in the solar system, including distance from the Sun, day length, atmosphere, and other structural characteristics.

2. Explain the theory scientists have developed about how the surface of Mercury was formed.

3. Discuss why Mercury is difficult to see.

Chapter 5 – Venus

Chapter Objective:

- Apply the basic astronomy concepts in relation to the planet Venus, and the important role Venus has in the solar system

The planet Venus is the second planet from our Sun and the closest to planet Earth. Because it is so close to Earth, Venus is one of the brightest objects in the night sky and is most visible just before dawn or just after sunset.

Venus is a terrestrial planet with a similar size, gravity and overall composition as Earth. Venus is 90% of Earth's diameter and has a mass that is 80% that of Earth. The atmospheric pressure, however, is 90 atmospheres at the surface, which is the equivalent of 1 kilometer underwater on Earth.

Space missions from the United States and Russia have sent probes to analyze the atmosphere of Venus to determine if the planet was suitable for human life. The characteristics of Venus (size, etc) indicated that it might be able to support life. Instead, the probes showed that Venus is extremely hot and dry, and has no signs of life. In fact, it is probably the least likely planet in the solar system to support life. The atmosphere of this planet is extremely dense and is comprised mainly of carbon dioxide with nitrogen and trace amounts of water vapor, acids, and heavy metals. Clouds in the atmosphere contain sulfuric acid.

Venus Image Courtesy of NASA

Why is Venus so hot? The heat from the Sun is able to penetrate the planet's atmosphere and reach the surface, and is prevented from escaping back into the atmosphere due to the carbon dioxide present in the atmosphere. This is a good example of what is known as the greenhouse effect, resulting in surface temperatures as high as 900 degrees Fahrenheit. These temperatures are even hotter than Mercury.

Venus has a very slow rotation, taking 243 Earth days for one rotation, or Venus day. This causes an interesting phenomenon of the same face of Venus facing the Earth when the planets are at their closest approach, or distance from one another. Venus has very strong winds in its upper atmosphere (350 kilometers/hours), with much slower winds at the surface that move at only a few kilometers per hour.

Venus has a topography that is quite simple – it is mostly rolling plains. There are also several broad depressions and some highlands. The surface is covered in volcanic flows resulting from volcanoes like the ones we have on Earth that occur as a result of tectonic plate activity. Astronomers also believe that there were oceans on the planet when it was much younger that have since boiled off because of the extreme heat.

Astronomers also think the internal composition of Venus is similar to that of Earth, with a solid inner core made of iron and surrounded by molten rock (magma). The crust is thought to be thicker than that on Earth and the pressure caused by convection within the magma is relieved in several small areas across the planet rather than the large volcanoes we experience on Earth.

.The Birth of Venus by Botticelli

Venus is named after the Roman goddess of love and beauty, probably because it is the brightest planet the ancient astronomers could identify. Following in this tradition, most of the features of the planet's surface are also named after females. For example, Aphrodite Terra (a highland area near the equator that is about the size of South America) and Guinivere Planitia (a broad depression).

Concept Reinforcement:

1. Describe Venus' position in the solar system, including distance from the Sun, day length, atmosphere, and other structural characteristics.

2. Explain why Venus is unique in the solar system.

3. Discuss why scientists thought Venus might support life in a similar manner to Earth.

Chapter 6 – Earth

Chapter Objective:

- Apply the basic astronomy concepts in relation to the planet Earth, and the important role Earth has in the solar system

The Earth is one of the smallest of the eight planets. Its diameter is approximately 7,900 miles, and its shape is described as being an oblate spheroid, which means that its equatorial diameter is longer than its polar diameter by about 27 miles.

The Earth was estimated to have formed about four and a half billion years ago, with the rest of the solar system. It was the discoverer of Halley's Comet who tried to estimate the age of the Earth by calculating the amount of salt the rivers had poured into the seas over time.

Earth as seen from Apollo 17 image courtesy of NASA

How are we able to sustain life on the planet? Earth's atmosphere is made up of 78 percent nitrogen, 21 percent oxygen, and one percent argon, with smaller quantities of water vapor, carbon dioxide, and other gases. The majority of the gases found in the Earth's atmosphere were probably released from underground volcanic activity.

The Earth is able to retain its atmosphere due to the interaction of the planet's mass and the gravitational pull. Earth's mass is great enough to keep most gases from escaping, except for the lighter gases hydrogen and helium. It is theorized that oxygen became part of the Earth's atmosphere when green plants came into existence, as plants produce oxygen by converting carbon dioxide through photosynthesis.

How was the atmosphere of the Earth formed? There are various hypotheses as to the origin of these gases. One theory states that the gases were trapped in layers of rock beneath the surface. These gases eventually escaped, mainly through volcanic eruptions, to form the atmosphere. It is also believed that comets passing through Earth's atmosphere deposited the elements found there. The proportion of carbon and nitrogen found in the atmosphere of the Earth is roughly the same as that found in the composition of comets.

The Earth has several layers that comprise the atmosphere, which changes in pressure and density with altitude. The atmosphere extends above the surface of the Earth many thousands of miles, but at least 95 percent of its total mass is found within 12 miles of the planets surface.

What are the different layers of the atmosphere, and what are they comprised of? There are six identifiable layers to the Earth's atmosphere: the troposphere, stratosphere, ozone layer, mesosphere, thermosphere, and the exosphere (ionosphere).

The bottom layer of the atmosphere is called the **troposphere**. This level contains the clouds and all the weather patterns. Temperatures drop rapidly at higher altitudes in the troposphere.

The next layer, the **stratosphere**, is about 9 miles above the surface of the Earth. The temperature is about -58 degrees Fahrenheit. There is a warm area about 25-45 miles, and that area is known as the ozone layer.

The **ozone layer** is considered the most important layer, as the oxygen layer in that region has three atoms per molecule, as opposed to the normal two. The ozone absorbs the ultraviolet rays from the sun, which heats up the space around it. The ozone layer also protects us from the harmful effects of the sun's rays. In recent years, the ozone layer has been damaged by chemical substances that we produce, such as CFCs (chlorofluoro-carbons). These substances are now banned in most countries, but were commonly found in the refrigerant of our refrigerators and also in hairsprays.

The region of the stratosphere above the ozone layer is the **mesosphere**, which is about 40 -50 miles above the surface of the Earth. The temperature is the same as that of the area just below the ozone.

The next layer is the **thermosphere**, which extends from 50-200 miles above the surface of the Earth. The temperature in this region of the atmosphere is about 1,800 degrees Fahrenheit (982 °C).

The **exosphere** is the highest layer of the Earth's atmosphere, which at its lowest point is about 200 miles above the surface of the Earth. It is within this layer that the molecules of gas found within break down into atoms. Many of the atoms become ionized or electrically charged by the sun's rays. This upper layer of the Earth's atmosphere is also known as the ionosphere due to this phenomenon.

Earth's Atmospheric Levels

The atmosphere of the Sun is unique in comparison to other planets in our solar system. It is the only known planetary atmosphere to sustain life. The moon and the planet Mercury have no atmosphere, in comparison. The atmospheres of Jupiter, Neptune, and Saturn are much more massive than the planet Earth. The atmosphere of the Earth is comprised mainly of nitrogen, while the atmospheres of Mars and Venus are comprised mainly of carbon dioxide.

Earth is unique in other respects as well. The Earth has a magnetic field that is probably the result of heat and motion in the Earth's core, which contains liquid metal.

The rotation of the Earth causes the core to act like a giant electrical generator, which cerates electricity and magnetism. This magnetic field extends far above the surface of the Earth, several times the radii into space. This area is known as the magnetosphere.

Concept Reinforcement:

1. Describe Earth's position in the solar system, including distance from the Sun, day length, atmosphere, and other structural characteristics.

2. Explain why Earth is unique in the solar system.

3. List and give a key characteristic of the Earth's atmospheric layers.

Chapter 7 – Mars

Chapter Objective:

- Apply the basic astronomy concepts in relation to the planet Mars, and the important role Mars has in the solar system

Mars is the fourth planet from the Sun in our solar system. Mars is about half the size of Earth, and has a rotation period which is a little longer than that of one Earth day. It takes Mars 687 Earth days to orbit the Sun, so the seasons are about twice as long as ours. There are two polar caps on Mars, with the northern cap larger and colder than the southern one. Mars has two small moons, Deimos and Phobos, which orbit the planet. Mars is the only planet whose surface can be clearly seen from Earth.

Mars image courtesy of NASA

Mars was named for the Roman god of war. It was long believed that Mars had life, possibly intelligent and more advanced than humans. With space probes launched by the US and the former Soviet Union in the 1960s and 1970s, these speculations that there was life on Mars were put to an end.

The spacecraft sent to Mars showed that the planet is a barren, desolate, crater covered environment, exposed to frequent, violent dust storms. There was no liquid water and little oxygen. The levels of ultraviolet radiation are strong enough to destroy any life as we know it. The temperature on Mars varies between -20 degrees Fahrenheit during the day and as low as -120 degrees Fahrenheit in the evening.

Mars has some notable topography, which includes a 15 mile high volcano called Olympus Mons, which is the largest in the solar system, and a 2,000 mile long canyon called Valle Marineris. This canyon is 26 times the length and three times as deep as the Grand Canyon.

There has been much speculation as to the possibility of life on Mars. Both the former Soviet Union and the United States led space missions to Mars. The Soviet Union was the first nation to send an unpiloted mission to Mars, which was unsuccessful. In 1971, Mariner 9 from the United States became the first spacecraft to orbit Mars. Mariner 9 sent back photos of a severe Martian dust storm, about 90 percent of the planet's surface, and the two moons of Mars. It showed that the northern hemisphere of the planet showed a younger surface appearance than the older cratered looking surface of the southern hemisphere.

Mars Surface as taken by the Pathfinder image courtesy of NASA

The first successful soft landing on Mars was made on July 20, 1976, by U.S. probes Viking 1 and Viking 2, each of which had an orbiter and a lander. Photos and soil samples were taken from either side of the planet, showing rust colored rocks and boulders with a reddish sky in the background. The presence of iron oxide in the soil is illustrated by the rust color. The soil samples confirmed that there was no apparent life on Mars.

In August 1996, a team of NASA led researchers re-analyzed a 4.5 billion year old rock that was found in Antarctica, identified as a fragment of the asteroid Vesta. The analysis showed that the chemical composition matched the soil on the surface of Mars. This fragment, known as ALH 84001, has been studied ever since, and recently discovered that the there are tiny, sausage shape markings, which do resemble the fossilized bacteria found in rocks on Earth. This may be the earliest form of life on Mars.

There are future exploration trips planned for Mars, which are expected to bring back rock and soil samples for further scientific examination.

Concept Reinforcement:

1. Describe Mars' position in the solar system, including distance from the Sun, day length, atmosphere, and other structural characteristics.

2. Explain the significance of the rock sample called "Vesta."

3. Discuss the various missions to Mars and their purposes.

Chapter 8 – Jupiter

Chapter Objective:

- Apply the basic astronomy concepts in relation to the planet Jupiter, and the important role Jupiter has in the solar system

Jupiter is the largest planet in our solar system. It is the fifth planet out from the Sun, and thirteen hundred times larger than Earth. The mass of Jupiter is about 300 times that of Earth. The diameter of Jupiter is 85,000 miles across, compared to 7,900 miles, the diameter of Earth.

Jupiter has sixteen moons, and has been considered its own mini solar system. It is the brightest object in the sky after the sun and Venus.

Jupiter's rotation is about 10 hours, compared to Earth's 24 hours.

Jupiter photo courtesy of NASA

Astronomers believe that the core of Jupiter is composed of a rocky material similar to Earth's, but has a diameter of at least 10 times that of Earth's core. The temperature of the core of Jupiter may be as hot as 18,000 degrees Fahrenheit, with the pressures at least two million times those at Earth's surface. Jupiter also has an intense magnetic field, which is five times that of the Sun's. Scientists believe that this is due to a layer of compressed hydrogen which surrounds the core, and acts like metal.

Jupiter has a unique feature which makes it stand out. The Great Red Spot is a swirling windy storm over 8,500 miles wide and 16,000 miles long, which is an area that can cover two Earths. Winds blow counterclockwise around the Great Red Spot at about 250 miles per hour. The color that is visible may be due to the sulfur or phosphorous. Beneath the red spot are three white oval areas, each of which is about the size of Mars.

There are a couple of theories regarding the origin of Jupiter. One states that Jupiter is comprised of the original gas and dust that came together to form the Sun and the planets. Another theory states that perhaps Jupiter was formed from the ice and rocks from comets, and it grew in size by attracting other particles around it.

Jupiter has been observed since the beginning of recorded time. Galileo first observed Jupiter and four of its moons through a homemade telescope. Each of the four moons is quite unique as their landscape has been formed by their gravitational pull in relation to Jupiter.

The four Galilean moons are: Ganymede, which is the largest and has its own magnetic field and a thin atmosphere; Io, which is the closest to Jupiter and the most volcanically active object in the solar system, due to the gravitational pull from Jupiter and the other moons; Callisto, which is the farthest moon that is characterized by craters, and Europa, which is covered with ice. There are a total of 63 moons around Jupiter. This is the most moons of any planet in the solar system.

Jupiter's Moons image courtesy of NASA

Astronomers first calculated the speed of light by studying the interactions of Jupiter's moons, in relation to the orbits of Earth and Jupiter. In 1676, Olaus Roemer calculated the speed of light based on initial calculations of astronomer Gian Domenico.

Jupiter is a unique planet in our solar system, as scientists believe it may hold information about the creation of our universe.

The space probe Galileo uncovered some interesting facts about Jupiter and its moons. It discovered a belt of radiation about 31,000 miles above Jupiter's clouds, which contain the strongest radio waves in the solar system. In addition, it found that Jupiter's clouds contain water, helium, hydrogen, carbon, sulfur, and neon in much smaller quantities than expected. It also discovered greater amounts of gaseous krypton and xenon than originally estimated.

No water was uncovered by Galileo's mini-probe, which indicates that the presence of life is unlikely. However, Jupiter's moon Europa, which is covered in ice, may have an ocean, possibly 60 miles deep. This ocean could be kept in the liquid state by the force of tidal heating. The possibility of an ocean suggests that there could be life.

Concept Reinforcement:

1. Describe Jupiter's position in the solar system, including distance from the Sun, day length, atmosphere, and other structural characteristics.

2. Explain why Jupiter is unique in the solar system.

3. Discuss the importance of Jupiter's moons to the study of astronomy.

Chapter 9 – Saturn

Chapter Objective:

- Apply the basic astronomy concepts in relation to the planet Saturn, and the important role Saturn has in the solar system

Saturn is the sixth planet from our Sun and is the second largest planet, after Jupiter. It is the only planet with a density that is about thirty percent less than water. To illustrate how light Saturn is: if the planet was placed in a large ocean, it would float.

Saturn image courtesy of NASA

Saturn is over 9 times wider and 95 times larger than Earth. It is also farther from the Sun than Earth. Saturn rotates very quickly on its axis, as it only takes approximately ten hours and thirty-nine minutes for one rotation. Due to this rapid rotation Saturn is flattened at its poles. The diameter at the equator of Saturn is about 10% larger than the diameter at its poles. Saturn's orbit around the sun is much longer than that of the Earth, as it takes about twenty-nine and a half Earth years to complete an orbit.

Saturn is made primarily of gas. The hazy yellow clouds are comprised mainly of ammonia and are swirled into bands by strong winds, which travel as fast as a thousand miles per hour. Saturn's surface is covered by a sea of liquid helium and hydrogen, which eventually becomes a metallic form of hydrogen. The helium and liquid hydrogen produce strong electric currents that generate the planet's magnetic field. The core of Saturn is made up of rock and ice. The planet's atmosphere is made up of about 97 percent hydrogen, 3 percent helium, and trace amounts of methane and ammonia.

What are some unique features about Saturn?

About every thirty Earth years, following Saturn's summer, a large storm occurs. "The Great White Spot" is visible for almost a month, which looks like a spot light. The spot will then dissipate and stretch around the planet as a thick white stripe. The storm could be the result of the warming of the atmosphere, which will then cause ammonia to bubble up and solidify, and be disrupted by the planet's strong winds.

Perhaps one of the most striking features of Saturn is the characteristic set of rings. The three largest planets, Jupiter, Uranus, and Neptune also have rings, but not as striking as Saturn's.

Christian Huygens

Saturn's rings were originally thought to have been moons encircling the planet, based on observations by the astronomer, physicist, mathematician Christian Huygens. It was not until the Voyager space probe missions of 1980 and 1981 that over a thousand ringlets encircling Saturn were discovered encircling the planet at about 50,000 miles from the surface.

The rings of Saturn are about one mile thick and are divided into three main parts, the brighter A and B rings and the dimmer C ring.

The complete make-up of Saturn's rings is not entirely known, but they do contain dust and water. The water is frozen in various forms; varying from snowflakes to icebergs in diameter.

Concept Reinforcement:

1. Describe Saturn's position in the solar system, including distance from the Sun, day length, atmosphere, and other structural characteristics.

2. Explain the key characteristics of Saturn's rings.

3. Define the great white spot.

Chapter 10 – Uranus

Chapter Objective:

- Apply the basic astronomy concepts in relation to the planet Uranus, and the important role Uranus has in the solar system

Uranus is the seventh planet from our Sun and one of the "gas giants." It is four planets away from Earth and 1.78 billion miles from the sun, which is about twice as far from the Sun than its neighbor Saturn. Uranus is the third largest planet by diameter. Uranus' diameter is larger than that of Neptune, but its mass is less.

Uranus image courtesy of NASA

With the discovery of Uranus in 1781 by Sir William Herschel, the size of the known universe doubled. Uranus is 31,800 miles in diameter at the equator making it the third largest planet in the solar system. It is four times the size of Earth.

Uranus is a gas planet that consists mainly of rock and ice, with an atmosphere of about 83% hydrogen, 15% helium and 2% methane. Its cloudy atmosphere is composed of 83 percent hydrogen, 15 percent helium, and smaller amounts of methane and hydrocarbons. Uranus actually has the coldest atmosphere in our solar system, with a minimum temperature of 49K (-224°C). Wind speeds can be has high as 250 meters per second.

The color that Uranus gets is from the atmospheric methane absorbing light at the red end of the visible spectrum and reflecting light at the blue end of the spectrum. In the core of the planet is a mix of ice, ammonia, and methane surrounded by a rocky core. Uranus has 27 known moons.

Uranus has a magnetic field. It is different than most other planets because it is not centered in the center of the planet. Uranus is unique in our solar system because its axis of rotation is tilted sideways, almost to the point of the axis of rotation lining up with its orbital plane (the ecliptic plane) around the Sun.

Uranus was the Greek god of the heavens and considered the earliest supreme god. He was the father of Cronus (Saturn). He also fathered the Cyclops and the Titans, who were ancestors of the Olympian Gods (Zeus, for example).

Concept Reinforcement:

1. Describe Uranus' position in the solar system, including distance from the Sun, day length, atmosphere, and other structural characteristics.

2. Explain why Uranus is unique in the solar system.

3. What Greek deity was Uranus named after?

Chapter 11 – Neptune

Chapter Objective:

- Apply the basic astronomy concepts in relation to the planet Neptune and the important role Neptune has in the solar system

The planet Neptune is about 17 times more massive than Earth. It is the eighth planet from the Sun at 2.8 billion miles away, making it the most distant of the large planets. It is a billion miles out from Uranus. Neptune's day is about 16 hours and orbits the sun once every 165 Earth years. Neptune is the color of water and was named after the Roman god of the sea. Methane gas is responsible for the color blue that is seen from space. Neptune's outer layer is composed of hydrogen, helium, and methane at a temperature of -352 degrees Fahrenheit. There is then a layer of ionized water, ammonia, and methane ice, and within that is a rocky iron core.

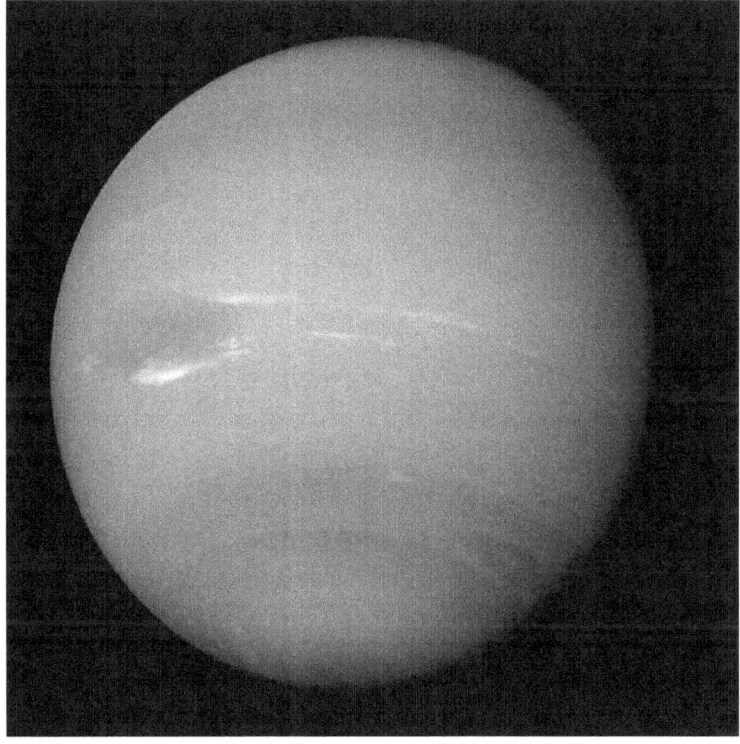

Neptune image courtesy of NASA

Neptune has the fiercest winds in the universe, with three storm systems apparent on the surface. The clouds on the surface of the planet swirl with the upper layer of methane crystals and rotate with the planet.

The most noticeable of the storms is a dark blue area, called the Great Dark Spot, which is as large as the Earth. The small dark spot is a storm area that is as large as our moon. There is also a small fast moving white storm system that follows the other storms on the planet.

Neptune has a magnetic field that is generated by pressurized hot water beneath the surface that reaches temperatures of 4,000 degrees Fahrenheit.

It wasn't until 1989 that we had the first pictures of Neptune when Voyager 2 collected some photos of the planet. The probe discovered that the planet is surrounded by five faint rings that are made up of small particles. It was further investigated and decided upon that these rings were really only arcs, not complete rings.

Gerard P. Kuiper

Neptune has thirteen known moons, with the largest being Triton, named for the mythical son of Neptune. Triton is 1,681 miles in diameter and orbits Neptune in a reversed orbit compared with the other satellites of Neptune, and also rotates in an opposite direction than that of Neptune. The moon Triton is considered the coldest place in the universe. The atmosphere of Triton is comprised of wind streaks, clouds, and haze, which has led to the belief that Neptune captured triton from an independent orbit around the sun. Nereid, which is Neptune's second moon, was discovered in 1949, by the Dutch astronomer Gerard P. Kuiper. There are six other moons that range in size from 31 miles in diameter to 250 miles in diameter.

Concept Reinforcement:

1. Describe Neptune's position in the solar system, including distance from the Sun, day length, atmosphere, and other structural characteristics.

2. Discuss the storms on the surface of Neptune.

3. Detail Neptune's moons.

Chapter 12 – Pluto

Chapter Objective:

- Apply the basic astronomy concepts in relation to Pluto, and the important role Pluto has in the solar system

Pluto was discovered in 1930 and was originally classified as the ninth planet in the solar system. It wasn't until 2006 that Pluto was reclassified as a dwarf planet. A dwarf planet is a celestial body orbiting the Sun, large enough to be rounded by gravitation pull, but not a satellite. Currently, the International Astronomical Union (IAU) recognizes only 3 dwarf planets; Pluto, Ceres, and Eris.

Artists concept of Pluto as seen from a nearby moon image courtesy of NASA

Pluto has been at the center of the debate as to its classification for several decades. Until recently, it was the smallest planet in the solar system and travels on an inclined orbit that actually crosses the plane of all the other planetary orbits. It will also cross over the orbital plane of its closest neighbor, Neptune. Additionally, Pluto does not fit the description of other orbiting bodies around the Sun; namely comets and asteroids.

Pluto was discovered by the American astronomer Clyde Tombaugh when looking for a planet called Planet X, which was causing disturbances in the orbit of Uranus. The gravitational field of Neptune accounted for some of the orbital irregularities, but in the end the ultimate result was due to Planet X, or Pluto.

The search initially began by looking at photographic plates, which were the road maps to the sky. On February 18, 1930, Tombaugh discovered Pluto, which was originally photographed by Percival Lowell, who was the founder of Lowell Observatory in Flagstaff, AZ.

The name of Pluto (the god of the underworld, in Greek mythology) was given to the then called ninth planet, for several reasons. Pluto is the farthest from the Sun and almost always dark. The sunlight that Pluto receives is about equal to the level of moonlight that Earth receives. Pluto is also the mythical brother of Jupiter and Neptune. The planet Pluto begins with the letters PL and those are the initials of Percival Lowell, the astronomer who has initially searched for the existence of the ninth planet (Planet X).

Since Pluto is so far from the Sun, it takes almost 250 years to complete one revolution around the sun. The length of a day on Pluto is 6.39 times longer than that of Earth's.

Pluto's moon, Charon, was discovered in 1978 when both Pluto and Charon were observed moving together into the inner solar system. The two bodies actually eclipsed each other, which allowed astronomers to plot the observed brightness curves of the planets. Initially, it was believed that Pluto and its moon were one large object. The moon Charon is a little more than half that of Pluto, which makes it the largest moon in relation to its planet in the solar system. The two bodies have been considered a double-planet in the past.

Pluto is a small dwarf planet, only 1,457 miles across, which is only 18 percent of Earth's diameter. It was originally thought that Pluto was a large planet, as its gravitational pull affected that of Uranus', which is two planets away. At the time, Pluto disrupted the theory that all the dense planets were closest to the sun and the giant gaseous planets were the farthest away.

As of today, Pluto remains unvisited by any NASA space probes, although a probe is expected to reach Pluto by 2015. Photos taken by the Hubble Space Telescope (HST) in 1996 showed that Pluto has frozen gases, polar caps, and light and dark spots. It is believed that Pluto is composed mainly of rock and ice, with surface temperatures between -350 and -380 degrees Fahrenheit (-212 and -229 degrees Celsius). The very bright areas on the surface are most likely nitrogen ice, solid methane, and carbon monoxide. The dark spots may be hydrocarbons resulting from the chemical activity of methane gases. The atmosphere is most likely comprised of nitrogen, carbon monoxide, and methane. When Pluto's orbit is closest to the Sun, the atmosphere is in the gaseous state, while the remainder of the atmosphere is frozen.

Many of the theories about the origin of Pluto connect Pluto with Neptune's moon Triton. Pluto rotates in a direction opposite that of most other planets and their satellites. One theory states that Titan, Pluto, and Charon are the only remaining bodies in our solar system which drifted into the Oort cloud; which is the area surrounding the solar system, where comets develop. Another idea states that Pluto could be one of Neptune's moons, and was struck by a very large object, breaking Pluto in two to form the moon and Pluto, and sent it into orbit around the sun. One more popular idea states that Pluto and Triton started out in independent orbits and Triton was captured by Neptune's gravitational field.

Concept Reinforcement:

1. Describe Pluto's position in the solar system, including distance from the Sun, day length, atmosphere, and other structural characteristics.

2. Discuss why Pluto is no longer considered a planet in the solar system.

3. Explain the theories about the origin of Pluto.

Chapter 13 – The Kuiper Belt

Chapter Objective:

- Apply the basic astronomy concepts of the Kuiper Belt, and the role the Kuiper Belt has in the solar system.

What is the Kuiper Belt, and why are astronomers interested in studying its origin? The Kuiper Belt is a flat disk of rings circling the solar system. It is located past the orbit of Neptune and is most likely comprised of planetary fragments and icy bodies which lie on the same plane as the planetary orbits. The Kuiper belt is between 30 and 50 astronomical units (AU) from the sun. 30 AU is about the distance to Neptune and 50 AU is about 7.5 billion km (4.7 billion miles) from the sun. The Kuiper Belt is thought to be the source of short period comets, which are comets that have periods (orbits) of less than 200 years.

The Dutch astronomer Gerard Kuiper formulated the theory about the probable make-up of the Kuiper Belt and theorized that the planetary matter found never formed an actual planet as the material was spread so thin and the pieces rarely collided with one another. The Kuiper Belt is also believed to contain material left over from the formation of the solar system.

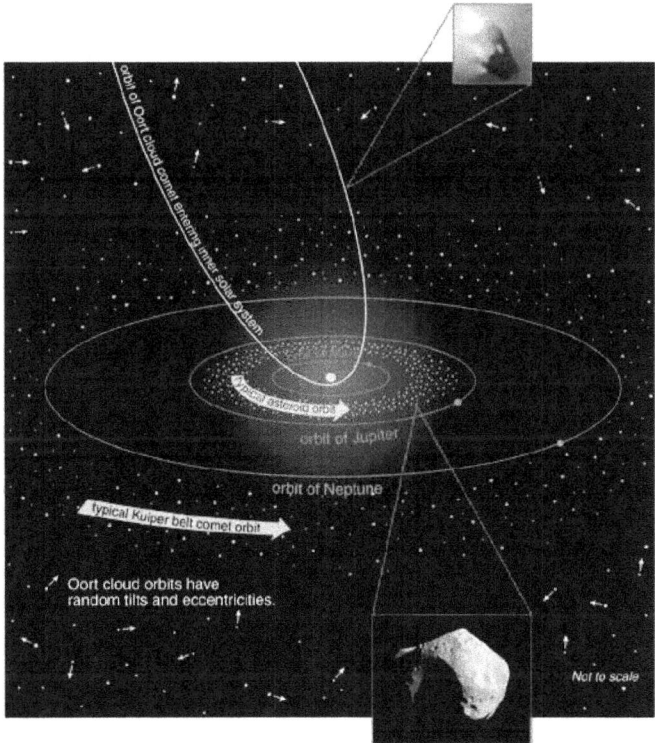

The Kuiper Belt image courtesy of NASA

Astronomers currently estimate that there are more than 35,000 objects in the Kuiper Belt that are more than 100 km in diameter. The Hubble Space Telescope allowed astronomers to detect even smaller objects in the Kuiper belt, estimated at approximately 20 km in diameter. There may be as many as 100 million of these small objects in the Kuiper Belt.

Credit NASA

Is the Kuiper Belt visible from Earth? The material in the Kuiper belt has only recently been detected using sophisticated telescopes. The first discovery was made in 1992, by David C. Jewitt and Jane Luu at the Mauna Kea Observatory in Hawaii. The Kuiper Belt was found beyond Pluto and is about 120 miles across and orbits the Sun from 3.2 billion miles away. There have been many more discoveries since then and there will be thousands more made in the future as telescopes on Earth become more powerful.

The Hubble Space Telescope has provided evidence to support the existence of the Kuiper belt beyond the planet Neptune. It has also been speculated that Pluto and its moon Charon are members of the Kuiper belt.

Concept Reinforcement:

1. Describe the Kuiper Belt.

2. Discuss the theory that Gerard Kuiper formulated.

3. Explain the recent detection of the Kuiper Belt.

Chapter 14 – Comets

Chapter Objective:

- Apply the basic astronomy concepts of comets and the role comets have in the solar system

What is a comet? Comets have often been described as dirty snowballs, which are actually made up of rocky material, dust, frozen methane, ammonia, and water. Comets travel in long elliptical orbits around the sun. A comet looks like a star and its structure consists of a nucleus, a head, and a gaseous tail. The tail of a comet is formed when some of the comet melts as it nears the sun. The tail always points away from the sun due to the solar wind (electrically charged subatomic particles that flow from the Sun).

Halley Comet courtesy of NASA

Comets are considered members of our solar system as they orbit the sun. It was believed that there were two groups of comets and the one group was thought to have followed a parabolic path, as it appeared only once. Further studies showed that all comets follow elliptical paths, but some comets' paths are so elongated that it could take millions of years for them to complete their paths, thus making it seem like their paths are parabolic.

How and where do comets originate? The Dutch astronomer Jan Hendrick Oort developed the commonly accepted theory about the origin of comets. It states that there are trillions of inactive comets that lie at the outer reaches of the solar system, about a light year from the sun. They remain in what is called the Oort cloud, a spherical region of space that encapsulates our solar system. The Oort cloud is far beyond the orbit of Pluto, about one to two light years from the sun, and extends halfway to the closest star, Proxima Centauri.

Comets remain in the Oort cloud until a passing gas cloud or star forces them into orbit around the sun. Another Dutch astronomer, Gerard Kuiper, suggested that there is another location for comets just beyond the edge of our solar system, about a thousand times closer to our sun than the Oort cloud. The Kuiper belt is between 35 and 1,000 AU (astronomical units – 1 AU is about 150 million km) from the sun, and contains ten million to one billion comets. This number is far fewer than the number of comets to be found in the Oort cloud.

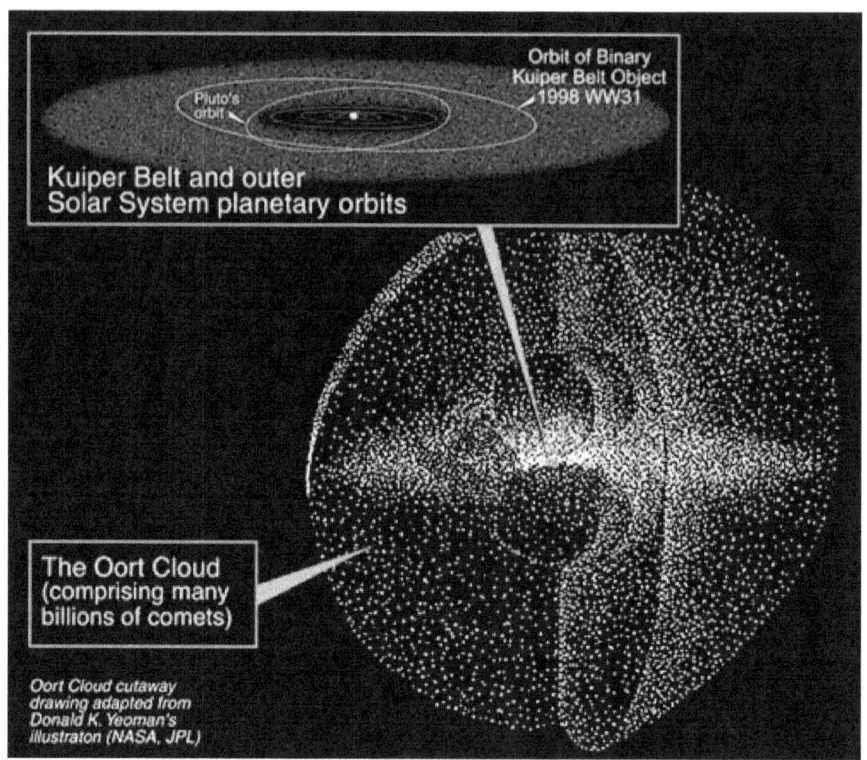

Artists rendering of the Kuiper Belt and Oort Cloud. Image courtesy of NASA

What happens at the end of a comet's life? There are several theories that discuss this topic, with one being that the comet's nucleus either will split or explode, which may produce a meteor shower. It is also speculated that the comet will eventually become inactive and become an asteroid. Comets also may be affected by gravity or other spatial disruptions, which will force the comet to exit the solar system and travel out into interstellar space.

Concept Reinforcement:

1. Describe a comet.

2. Discuss the impact of Jan Hendrick Oort on astronomy.

3. Explain three theories of how comets die.

Chapter 15 – Past, Present, and Future Space Travel

Chapter Objective:

- Explain past, present and future space travel

The idea of space travel has been with us for at least as long as we have gazed up to the skies. We have researched and explored the positions, motion, composition, energy, and evolution of celestial bodies and other, sometimes unexplainable phenomena such as the Sun, meteors, comets, stars, and the planets along with their moons and satellites.

When man first landed on the moon in 1969, people were able to vicariously experience space travel. Today there is an endless amount of information at our fingertips when it comes to the internet. We have access to information on past, present, and future space travel.

Apollo 12 astronaut Alan Bean climbs down the lunar module Intrepid, joining Pete Conrad on the Moon 35 years ago — November 19, 1969.

The second lunar landing mission, Apollo 12 proved the astronauts could make a precise landing. It also gave the crew a chance for a unique rendezvous with the robotic explorer Surveyor 3, which had been on the Moon since 1967. Conrad and Bean spent more than 31 hours on the surface before rejoining crewmate Dick Gordon orbiting overhead in the command module Yankee Clipper. Credit: NASA

Space travel has been influenced in many ways; socially, politically, economically, and from a military perspective, as well. The imaginations of scientists, engineers, and the general public have been influenced in some way as to the topic of space travel.

There were many reasons for exploring space through space travel, including advancing scientific research, uniting different nations, ensuring the survival of the human race, and developing military/strategic advantages against other countries.

In the early years of space exploration, there was a "Space Race" between then called Soviet Union and the United States. The launch of the first man made object into space to orbit the Earth, was the USSR's Sputnik 1, on October 4, 1957. The first Moon landing by the American Apollo 11 craft, on July 20, 1969, was considered the milestone in the so-called space race. The very first man made objects to reach space were Nazi-Germany's V2 rockets that were used as early as World War II.

After the first 20 years of space exploration the focus changed to more long standing programs, such as the Space Shuttle program, and the International Space Station, which proved that the world went from a more competitive to cooperative space program.

The Way Home

Backdropped by a cloud-covered part of Earth, Space Shuttle Atlantis was photographed by the Expedition 15 crew after it undocked from the International Space Station on June 19, 2007, in preparation for the journey home. The STS-117 astronauts completed about eight days of joint operations with the station crew. The docked Soyuz spacecraft is visible at left. Image credit: NASA

Today there are about seventy-three space missions that are either funded by a single country or as part of joint projects by more than one nation. These space missions are a combination of orbital and planetary missions, which involve getting samples not only from space, but from Earth and other planets as well.

In the 1990s private interests started helping private individuals develop space tourism. The idea of space tourism is a recent venture that was pioneered by Russia. Tourists actually pay for flight into space. Orbital space tourism is very limited and expensive. This is only available through the Russian Space Agency.

These space flights are a thrill seeking adventure that can cost about twenty million dollars. Those who go on these trips are awed by the experience of looking at Earth from space and the feeling of weightlessness. For some individuals it is the idea of status that is achieved from this adventure.

The most notable space project to date is the Hubble Space Telescope which was launched in 1990 and is expected to be in orbit until 2010. It will be on display at the Smithsonian Institution in Washington, DC, after its mission is complete. There are larger government programs, which have advocated manned missions to the Moon and possibly Mars after 2010.

Caption: In one of the most detailed astronomical images ever produced, NASA's Hubble Space Telescope captured an unprecedented look at the Orion Nebula. This turbulent star formation region is one of astronomy's most dramatic and photogenic celestial objects. More than 3,000 stars of various sizes appear in this image. Some of them have never been seen in visible light. These stars reside in a dramatic dust-and-gas landscape of plateaus, mountains, and valleys that are reminiscent of the Grand Canyon. The Orion Nebula is a picture book of star formation, from the massive, young stars that are shaping the nebula to the pillars of dense gas that may be the homes of budding stars.

Image credit: NASA, ESA, M. Robberto (Space Telescope Science Institute/ESA) and the Hubble Space Telescope Orion Treasury Project Team

Re-deploy: Hubble is leaving the payload bay of the Shuttle with Earth in the background
Image credit: NASA

The future of space travel is unknown at this point, although science fiction has proposed ideas such as suspended animation for long space flights, wormholes that allow rapid jumps from one part of the universe to another, transporter technology that allows objects and bodies to be beamed through space, and so on. The more likely outcome is that humans will explore the planets of our solar system more thoroughly using manned and unmanned probes, as well as increasing the ability of people to live in space stations and extract natural resources from the moon and other planets.

Concept Reinforcement:

1. Describe the space race.

2. Explain space tourism.

3. Discuss the importance of space exploration and how the Hubble Telescope is supporting this mission.

4. Explain the difference between science fiction space travel and what is more likely to happen as humans develop the tools to explore space more thoroughly.

Chapter 16 – The Galaxy

Chapter Objective:

- Introduce the basic principals of astronomy in relation to the galaxy

What is a galaxy? The galaxy is a region of space that is made up of hundreds of billions of stars, planets, nebulae, dust, gas, and empty space. A black hole, which is a single point of infinite mass and gravity, may also be found at the center of the universe. There are at least 50 billion galaxies believed to contain most of the observable mass in the universe. There is also invisible dark matter, which may make up about 90 percent of the mass in the universe.

Galaxies can be different shapes; elliptical, spiral, or irregular in shape. The spiral shaped galaxies that are nearby are Andromeda and the Milky Way. The objects are centered (stars and a black hole) and surrounded by an invisible cloud of dark matter, and looks like a pinwheel, that is made up of stars on the outer edges. The entire galaxy is rotating, and the stars on the outer edges make up the arms.

The milky way galaxy image courtesy of NASA

An elliptical galaxy contains mainly older stars that contain very little dust or gas. The shape of an elliptical galaxy is either round or oval, and flattened or spherical. It looks like the nucleus of a spiral galaxy without the arms. It is of interest to point out that spiral galaxies can lose their arms, but elliptical galaxies do not form arms. Approximately a quarter of all galaxies are oddly shaped. There are fast moving spirals, ring galaxies that appear to have no nucleus, and ribbon like galaxies, that appear twisted and form when two galaxies collide.

Andromeda galaxy image courtesy of John Lanoue

When did we first learn about the existence of other galaxies?

The American astronomer Edwin Powell Hubble first proved the existence of other galaxies in 1924. Hubble used a very powerful 100-inch telescope at Mount Wilson Observatory, where he observed a group of stars that he thought was part of the Milky Way, but it turned out to be a separate galaxy. This galaxy is now known as the Andromeda galaxy. Hubble also discovered numerous other spiral galaxies.

What galaxy does Earth's solar system belong to?

The Earth's solar system belongs to the Milky Way galaxy. The Milky Way galaxy is about eighty thousand light years across. The nucleus contains billions of old stars and possibly a black hole. Earth's solar system lies in the Orion arm of the galaxy, which is about twenty eight thousand miles from the center of the galaxy.

The Milky Way is part of a cluster of galaxies that is called the Local Group, and the Local Group is part of a local supercluster that includes many other clusters. These superclusters are separated by extremely large voids of space. There are very few galaxies in between these voids.

What have we learned in recent years about galaxies?

The Hubble Space Telescope sent back photographs of about fifteen hundred distant galaxies in the process of formation, which indicated that the number of galaxies in the universe was greater than previously thought. Astronomers estimated the number of galaxies to be fifty billion.

If we think about the relationship of the Earth's solar system to the galaxy, and the universe, we quickly realize that our part of the universe is very small in comparison. We can study the interaction of the different galaxies and their components and structures and have a limited understanding of the vastness of the universe. We are still in the early stages of space exploration and have much more information to gather and more questions to answer.

Concept Reinforcement:

1. Describe a galaxy.

2. Explain the different shapes a galaxy can have.

3. Describe where the Earth is located within the Milky Way,

Chapter 17 – The Tools of Astronomers

Chapter Objective:

- Understand and explore the tools that are used by astronomers

We have been fascinated with the skies since the dawn of time. Ancient civilizations looked to the sky and studied the stars and their interactions in order to better understand Earth's place in the universe. Early civilizations left behind architectural structures to indicate their interest in the skies. Stonehenge and the great pyramids of Egypt are the most well known of these structures. The great pyramids align with Orion's belt and other important stars. Stonehenge, located in Great Britain, is believed to have been constructed to follow the alignment of the sun and determine the timing of an eclipse and perhaps also as an indicator to gauge the change of seasons.

The Great Pyramids of Egypt

Stonehenge in Great Britain

Perhaps the most well known tool of an astronomer is the telescope. The telescope was invented by Italian astronomer Galileo Galilei in 1610. It was a simple creation that consisted of a cylinder with a lens at both ends. It provided a simple magnification of twenty times that of the naked eye. Galileo was able to observe the planets, stars, our galaxy, the Milky Way and numerous other objects in space. Galileo formulated that the Earth was not the center of the universe, but rather the sun and all the planets orbited around the Sun.

Galileo

The early telescopes had limited capabilities for observing the stars, as objects in the distant sky would be observed as fuzzy objects due to the interference from the atmosphere.

A simple telescope

It was not until Newton and Kepler formulated mathematical equations to explain the observations of astronomers, could their findings be accepted, which led to popular acceptance of the heliocentric model of the solar system. The heliocentric model was a fundamental change in how people viewed our universe. Prior to the heliocentric model being developed and proven, people thought the Earth was the center of the solar system and that everything revolved around our planet. This is known as the geocentric view of the solar system.

There are different types of telescopes used by astronomers. Reflecting and refracting telescopes are the most common types. These are more advanced telescopes than the simple telescope developed by Galileo. In order to get around the challenges presented by the distortion resulting from light coming through the atmosphere, large telescopes are placed at high elevations and even launched into space. The Hubble Space telescope is an example of such a telescope.

Concept Reinforcement:

1. Explain the purpose of Stonehenge in astronomy.

2. Describe a simple telescope.

3. Discuss why telescopes are placed at high elevations.

Chapter 18 – The Hubble Space Telescope

Chapter Objective:

- Understand and explore the Hubble Telescope and its impact on astronomy

The Hubble Space Telescope was a long time in the making. From the first concept by German rocketry specialist, Hermann Oberth, in 1923 to put a satellite in space using a rocket ship. In 1946, the American astrophysicist Lyman Spitzer, Jr. wrote a paper proposing a space observatory, and spent the next 50 years making the dream a reality.

The Hubble Space Telescope was named after the American astronomer Edwin Powell Hubble, who was the first to make the discovery that other galaxies even existed, and was also the first to make the distinction in the types of galaxies that exist. He made this discovery in 1924, observing a cluster of stars thought to be part of the Milky Way, but was actually a separate galaxy, now known as the Andromeda galaxy. Hubble postulated the theory that we live in an ever-expanding universe, which led support to the "Big Bang Theory", the theory that the Universe was created some 13.7 billion years ago, starting with an incredible explosion of matter.

The Hubble Space Telescope was first launched in April 1990, and has had 4 missions from Earth since that time. The telescope has revolutionized the field of Astronomy, as it has enabled astronomers to see photos of the Universe, from our solar system to galaxies far away.

The telescope was not perfect when it was first launched into orbit April 24, 1990. The first images that were sent back from the Hubble Space Telescope were blurry, and it was traced back to the telescope's primary mirror, which was about $1/50^{th}$ the thickness of a piece of paper, too thick.

It took scientists 11 months to train for the first ever repair in space. In December 1993, a crew of seven astronauts went to make repairs to the Hubble Space Telescope. The mission lasted five days and was successful. Scientists were able to make adjustments to the mirror with COSTAR, the Corrective Optics Space Telescope Axial Replacement, which could be installed in place of one of the telescope's other instruments. The COSTAR instrument would correct the images produced by the remaining and future instruments. Astronauts also replaced the Wide Field/Planetary Camera with a new version, the Wide Field and Planetary Camera 2 (WFPC2) that contained small mirrors to correct for the distorted images that were being transmitted.

The Hubble Space Telescope (HTS) begins its separation from Space Shuttle Discovery following its release on mission STS-82. Image courtesy of NASA

The structure of the telescope is about 43.5 ft long by about 14 ft wide, which is about the size of a tractor-trailer. It is constructed of a series of mirrors that reflect images to the computers on board. The computers can then transmit data images back to the space station headquarters at NASA. The Hubble Telescope circles the Earth, approximately every 97 minutes and is approximately 350 miles above the surface of the Earth. The scope travels at approximately 5 miles per second. The technology of the Hubble Space Telescope has provided a whole new perspective for gaining knowledge about the Universe as it transmits images from the formation of galaxies to remnants of supernovas.

What makes the technology of the Hubble Space Telescope so advanced?

The technology of the telescope is called a Cassegrain reflector, where mirrors work in concert with each other. The primary mirror reflects light to the secondary mirror, which then reflects the light through a small hole in the primary mirror and sends the data to the computers to be analyzed back on Earth. The image data the telescope transmits back to Earth is equivalent to about 120 Gigabytes, or over 3600 average size books.

We would not have the knowledge that we have today, if it were not for the capabilities of the Hubble Space Telescope. The images that the telescope can capture are superior to those from any telescope on Earth, as the telescopes on Earth still have atmospheric disturbance that blocks the transmission of light causes captured images to appear "fuzzy."

The Hubble Space Telescope is one of NASA's most successful space missions. It was started almost twenty years ago, and is still going strong. Each year scientists compete for the ability to utilize the telescope to capture images from different regions of space. There are over 1,000 proposals each year and only two hundred are awarded use of the telescope.

The Hubble Space Telescope has been one of the greatest resources to the NASA space program. It has contributed information about the early formation of galaxies, energy bursts from star collapses in distant galaxies, in addition to other phenomena. It is by far one of the most utilized observatories and has contributed to over 6,000 scientific articles since its launch into orbit.

Concept Reinforcement:

1. Explain the reason the Hubble Telescope was built.

2. Who was Edwin Powell Hubble?

3. Describe a Cassegrain reflector.

Chapter 19 – The International Space Station

Chapter Objective

- Understand and explore the International Space Station, and its impact on astronomy

The International Space Station (ISS) is a jointly sponsored research facility in space. The facility is being assembled in space with coordination from the space agencies of the United States (NASA), Russia (RKA), Japan (JAXA), and eleven European countries (ESA). The Brazilian Space Agency is a participant through a separate contract with NASA. There is also participation from Italy under a separate contract, as well.

The project began in 1998 and is anticipated to be completed in 2010. The space station will be in operation until 2016. The space station has been staffed continuously, with rotating astronauts from different countries, since 2000. The ISS is a continuation of other previously planned space stations, and is the largest station thus far. It is manned by three astronauts at a time but will eventually hold 6 crew members to achieve an active research program. Currently, the space station has been visited by astronauts from 16 countries. The ISS was also the destination for the first five "space tourists."

Backdropped by a blue and white part of Earth, the International Space Station is seen from Space Shuttle Discovery as the two spacecraft begin their relative separation. Image courtesy of NASA

Why makes the International Space Station such an important part of the International Space Program?

One of the most unique features of the ISS is the fact that it is being assembled entirely in Earth's orbit. The space station is visible to the naked eye, as it is in a low orbit about 217 miles above the surface, traveling at an average speed of just over 17,000 miles per hour. The ISS completes just over 15 orbits per day.

The space station project was first announced in 1993 and was called Space Station Alpha. It was initially planned to combine the space stations of all participating space agencies; NASA's Space Station Freedom, Russia's Mir-2 (after the Mir Space Station, whose core is now called Zvezda), and the European Space Agency's (ESA) Columbus, which was supposed to be a stand alone space lab.

As of July, 2008, the space station is 75% complete, and represents a major endeavor in aerospace engineering. The station utilizes the energy from the sun by converting it into electricity through the use of solar panels.

There are many advanced systems in place for all environmental systems, such as; life support and altitude control. The main function of the ISS is to conduct scientific research for experiments that will require one or more unusual conditions on the space station. There are several fields of research including biology (encompassing biomedical research and biotechnology), physics (encompassing fluid physics, materials science, and quantum physics), astronomy, and meteorology.

In 2005, the NASA Authorization Act designated the United States area of the ISS as a national laboratory with a goal in mind to increase the capabilities of the ISS by other Federal institutions and the private sector. So far, research into the long term effects of microgravity on humans is the only project underway. There are four new research modules set to arrive at the International Space Station by 2010, with more specialized research planned.

Future research will include the study of the effects of prolonged exposure to zero gravity on muscle atrophy, bone loss, and fluid shifts. This information will be utilized to evaluate space colonization and more lengthy space travel, as this may become more feasible.

The long term goals of the research are to develop the technologies that may become necessary for space and planetary exploration and colonization by humans. This would include research into life support systems and environmental monitoring–in short, a better understanding of the Universe.

Concept Reinforcement:

1. Describe the background of International Space Station.

2. Explain the importance of the International Space Station to the International Space Program.

3. Discuss some of the research goals of the ISS.

Chapter 20 – The Milky Way Galaxy

Chapter Objective:

- Apply the basic astronomy concepts to the structure of the Milky Way galaxy

The Milky Way Galaxy, home to our solar system, is a spiral galaxy with at least 200 -400 billion other stars, their planets, thousands of nebulae and clusters.

It is a very large galaxy with a mass between 750 billion and one trillion solar masses; its diameter is about 100,000 light years. The Hubble Space Telescope has revealed that the Milky Way is a spiral galaxy, based on radio astronomical investigations of the way hydrogen clouds are distributed. Our galaxy has a central disk like component with a spiral structure, as well as a well-defined nuclear region, which has a sizable halo-like component.

The Milky Way image courtesy of NASA

Why is our galaxy called the Milky Way?

The galaxy appears as a star filled expanse of light, stretching across the sky. In ancient times, it was believed that this glowing band resembled a river of milk, and became known as the Milky Way galaxy.

The Milky Way galaxy belongs to what is called the Local Group, which is a smaller group of 3 large and over 30 smaller galaxies, and is the second largest after the Andromeda Galaxy (M31), which is 2.9 million light years away. There are many galaxies that are nearby, but they are faint in comparison. There are also dwarf Local Group members which are considered satellites of the Milky Way.

The spiral arms of our Milky Way contain interstellar matter, young stars, open star clusters, and diffuse nebulae. The disk component (or center) consists mainly of old stars and contains the globular star clusters, of which 150 of the approximate 200 have been identified.

Our solar system lies in the outer regions of the galaxy. Our location is still within the disk, only about 20 light years above the equatorial symmetry plane, and 28,000 light years from the Galactic Center. Thus the Milky Way appears as a luminous band, which spins along this plane of symmetry, also known as the Galactic Equator. The center lies in the direction of the constellation Sagittarius, but also close to the constellations Scorpius and Ophiuchus.

Our solar system lies within a smaller spiral arm, called the Local or Orion Arm, which is a connection between the inner and outer arms; Sagittarius and Perseus.

Our sun, along with the whole Solar System, is orbiting the Galactic Center, in a nearly circular orbit. The speed at which our solar system orbits around the Galactic Center is approximately 250 km/sec. It takes approximately 220 million years to complete one orbit, which means that the Solar System has orbited the Galactic Center about 20 to 21 times since its formation about 4.6 billion years ago.

From the Earth, we can see only part of the Milky Way, as much of the galaxy is obstructed from view due to interstellar dust and gas. It is visible on clear summer nights, away from the glare of city lights.

Concept Reinforcement:

1. Describe the composition of the Milky Way.

2. Describe how the Milky Way appears to people on Earth.

3. Which arm of the Milky Way holds our solar system?

Chapter 21 – The Formation of the Milky Way Galaxy

Chapter Objective:

- Apply the basic astronomy concepts to the formation of the Milky Way galaxy

The Milky Way may not have formed the way astronomers previously thought, but by some other unknown process.

Image courtesy of NASA

The Milky Way galaxy is often referred to as "the galaxy." It is a barred spiral galaxy that is part of the Local Group of galaxies. It is only one of billions of galaxies in the observable universe. Our solar system is located within the Milky Way. The plane of the Milky Way galaxy is visible from Earth as a band of light in the night sky.

Milky Way image courtesy of NASA

It has been hypothesized that the Milky Way galaxy formed early in the life of our universe. As matter was quickly expanding outwards, clusters of large amounts of matter began to orbit around a common center mass. These were the early galaxies, which would grow in size from events such as collisions with other galaxies. So, the Milky Way galaxy is one of an unknown number of galaxies in the universe.

Milky Way Courtesy of NASA

The Milky Way galaxy is estimated to be about 13.2 billion years old, which is the estimated age of the oldest star in the galaxy. This star is almost as old as the universe, although it is impossible to determine its precise age at this point.

Concept Reinforcement:

1. Describe the Milky Way.

2. Explain how the Milky Way is thought to have formed.

3. How old is the Milky Way estimated to be?

Chapter 22 – Other Galaxies in the Universe

Chapter Objective:

- Understand and explore other visible galaxies

Galaxies vary in size and shape, as they can be spiral or elliptical, and other irregular shapes. The Andromeda and Milky Way galaxies are both spiral shaped.

These two galaxies are closely related, as they were both formed around the same time, have the same shape, and both are believed to have black holes at their center.

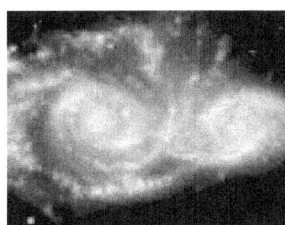  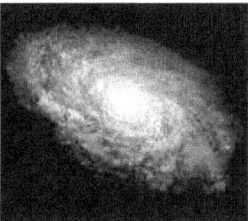

Images courtesy of NASA

The Andromeda galaxy is the closest major galaxy to the Milky Way at 2.2 million miles away. It is also the most distant object from Earth that is visible without the use of a telescope. The Andromeda galaxy is two times the mass of our own Milky Way galaxy.

The Andromeda galaxy contains hundreds of billions of stars, as some of them make up part of the six hundred or more globular clusters on the edge of the galaxy

Image courtesy of NASA

The two closest neighbors to the Milky Way are the Large and Small Magellanic Clouds, which are visible to the naked eye in the Southern Hemisphere. These two galaxies are fairly small and irregularly shaped. They were named after the explorer Ferdinand Magellan, who first recorded their existence in 1519.

The Large Magenellic cloud is about 163,000 light years from Earth, and the Small Magellanic cloud is about 195,000 light years from Earth.

What makes these two galaxies unique?

The Large and Small Magellanic clouds are unique as they house stars that are in their infancy. It is also a place where stars can be observed from their formation to their ultimate end in a supernova.

It was in fact in the Large Magellanic Cloud where the first supernova in over three centuries, occurred in 1987. It was the first time that a supernova was visible enough to be seen with the naked eye.

Concept Reinforcement:

1. Describe the Andromeda Galaxy.

2. Describe the Large and Small Magellanic Clouds.

3. Explain the shape of the Milky Way galaxy.

Chapter 23 – The Constellations

Chapter Objective:

- Understand and explore Constellations

What is a constellation?

A constellation is one of the eighty-eight groups of stars in the sky that were originally named for mythological beings. The constellations comprise the entire celestial sphere, which is the imaginary sphere that surrounds the Earth. Some of the constellations actually resemble the figures they were named after, and many were simply named in honor of them.

Scientists use the celestial sphere to plot the stars and other objects in space and can then chart their movement that is affected by the rotation of the Earth.

Star formation in the Orion constellation as photographed in infrared by NASA's Spitzer Space Telescope.

A print of the copperplate engraving for Johann Bayer's Uranometria showing the constellation Orion. This image is courtesy of the United States Naval Observatory Library

A constellation differs slightly from a star pattern or grouping, and such a pattern is called an asterism. An example of such a grouping is called the Big Dipper (so called in N. America) or the Plough (so called in the UK).

The big dipper

The stars in a constellation or an asterism will usually not have any astrophysical relationship to each other–they just happen to appear close to each other as viewed in the sky from Earth. They typically are many light years apart from each other. The exception to this is the Ursa Major moving group.

Dark cloud constellations or dark nebulae have been attributed to different cultures and their associations with seasonal or climatic changes. The Inca civilization identified several dark areas in the Milky Way as animals, and associated their appearance with the seasonal rains.

Chinese constellations are different from Western constellations, because of the development of ancient Chinese astronomy. The Western civilization has 12 zodiac constellations where the Chinese has what are known as Xiu or literally mansions. In Hindu or Vedic astronomy there are the 12 Rashi, or constellations, that correspond to the Western constellations, but these are further divided into 27 lunar houses, known as Nakshatras.

All the constellation names in the Western Hemisphere are derived from Latin and have a standard three letter abbreviation assigned by the International Astronomical Union, for example; Pisces (Psc) and Sagittarius (Sgr).

What is the zodiac in reference to the constellations?

The zodiac represents an annual cycle of twelve stations along the ecliptic, the path of the sun across the heavens through the known constellations. There are twelve equal zones that divide the ecliptic into zones of celestial longitude. The zodiac is the first known recognized celestial coordinate system.

Concept Reinforcement:

1. Explain what a constellation is.

2. Describe why constellations that appear to be near one another may not actually be close to another in space.

3. How does the International Astronomical Union standardize naming of western hemisphere constellations?

Chapter 24 – Stars

Chapter Objective:

- Understand and explore types of stars

What is a star?

A star is a massive ball of plasma, with the closest star to our Earth being the Sun. The Sun is the source of most of the energy on Earth.

What makes a star so visible in the night sky? Stars that are not outshone by the Sun are visible in the night sky. A star shines because of thermonuclear fusion that takes place in its core, which releases energy that transcends the star's interior and travels into outer space.

It is the process of nuclear fusion that enables a star to shine for most of its life. Almost all of the elements that are heavier than hydrogen and helium were created by fusion processes in stars.

The Pleiades, an open cluster of stars in the constellation of Taurus image courtesy of NASA

Astronomers are able to determine the mass, chemical composition, and age of a star by observing its spectrum, luminosity and its motion through space. It is the mass of a star that helps determine not only its evolution, but also help determine its fate. The other characteristics of a star are determined by its history, diameter, movement, rotation, and temperature.

The age and evolutionary state of a star can be determined by plotting their temperature against their luminosity. This plot is known as a Hertzsprung-Russell diagram.

The study of stars made rapid scientific advances in the 20th century. The photograph became a useful tool, in addition to the Hertzsprung-Russel diagram, for studying stars.

Scientists were able to develop models to explain the interiors of stars and stellar evolution. It was through the advances in quantum physics that scientists were able to explain the spectra of stars. This advancement also allowed for the determination of the composition of the stellar atmosphere.

Stars have had their own association with myths. Stars were thought to be souls of the dead or even gods.

Who can officially name stars?

The only body that is officially recognized by the scientific community as having the authority to name stars or other celestial bodies is the International Astronomical Union (IAU).

Where do stars form?

Stars primarily form within regions of higher density in the interstellar medium. These regions are called molecular clouds, and are comprised mainly of hydrogen, about 25% helium, and some heavier elements making up a smaller fraction. One example of a star forming region would be the Orion Nebula.

the Orion Nebula image courtesy of NASA

How are stars classified?

Early stars; those less than 2 solar masses are called T Tauri stars. Those stars with greater mass are known as Herbig Ae/Be stars. Stars that are newly formed will emit jets of gas along their axis of rotation (known as Herbig-Haro objects).

Concept Reinforcement:

1. Describe the physical characteristics of a star.

2. Explain how a star generates energy.

3. Describe a Hertzsprung-Russell diagram.

Chapter 25 – The Star Cycle

Chapter Objective:

- Understand and explore the star cycle

How does a star form?

A star begins as a collapsing cloud of material, which is comprised mainly of hydrogen, helium, and other trace heavier elements. Once the stellar core becomes sufficiently dense, some of the hydrogen will be steadily converted into helium through the process of nuclear fusion. The star's interior can then carry energy away from the center core through a combination of processes; both radioactive and convective.

It is the internal pressure of the star that prevents it from collapsing under its own gravity. When the hydrogen at the core is exhausted, stars having at least 0.4 times the mass of the Sun will expand to become a red giant, and may also fuse heavier elements at the core or around the core in one of the "shells." The star then evolves, recycling a portion of the matter into the interstellar environment. This matter will then form a new generation of stars. The proportion of heavy elements in these stars will be higher as well.

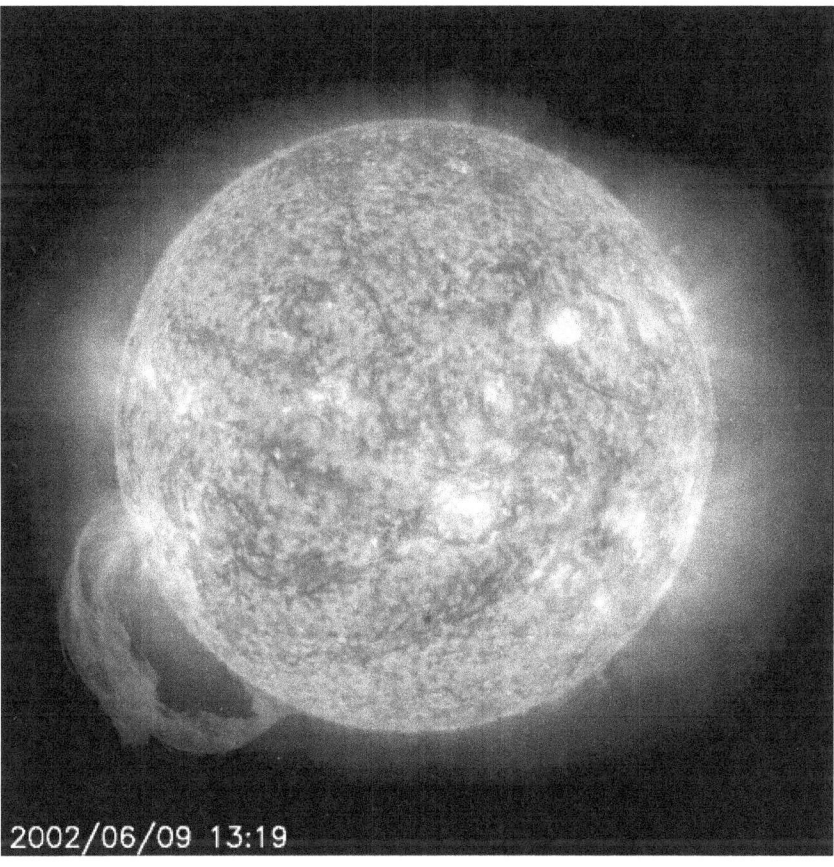

Red Giant. Image courtesy of NASA

Where do stars form?

Stars form from the gravitational instability inside a molecular cloud, which are often triggered by shockwaves from supernovae, which are massive stellar explosions, or by the collision of two galaxies. Once a region reaches a sufficient density of matter, it will begin to collapse under its own gravitational force. As the cloud collapses, the dense dust and gas form what are known as "Bok globules," containing up to 50 solar masses of material. As the globule begins to collapse and the density increases, gravitational energy is then converted to heat and the temperature rises.

A prostar forms at the core, when the protostellar cloud reaches a stable condition. This is followed by the gravitational contraction that actually forms the star, which will last approximately 10-15 million years.

Stars will spend about 90% of their lifespan fusing hydrogen to produce helium in high temperature and high-pressure reactions near their core. These stars are referred to as dwarf stars, and said to be on the main sequence. The proportion of helium in a star's core will steadily increase over time, and thus the star will increase in temperature and luminosity, in order to maintain the rate of nuclear fusion at the core.

The amount of time a star spends on the main sequence depends on the amount of fuel and its burn rate, which translates into how much mass and luminosity the star has.

Large stars burn their fuel rapidly and are short lived. Small stars also called red dwarfs, burn their fuel slowly and last billions of years. The stars will eventually become dimmer and dimmer.

Concept Reinforcement:

1. Explain how a star forms.

2. List the primary components of a star.

3. Describe the formation of a red giant.

Chapter 26 – Supernovas

Chapter Objective:

- Understand and explore supernovas

What is a supernova?

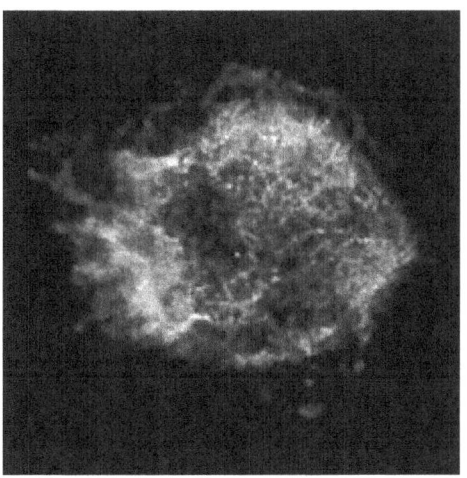

Image courtesy of NASA

A supernova is the end result of a star running out fuel. When a star at least eight times the mass of the sun runs out of the hydrogen, helium, and other trace elements that it needs to exist, it sheds much of its mass and becomes a supernova.

The supernova is a phenomenon that occurs only once in the life cycle of a relatively large star (those that are at least eight times the mass of the sun).

The star first collapses on itself, and then explodes outward with a huge force. The star sheds its outer atmospheric layers as a result of the explosion and will shine more brightly than the rest of the stars in the galaxy all together.

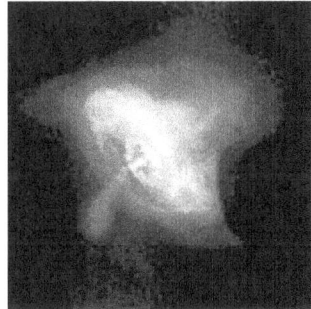

Image courtesy of NASA

What happens to the star after the supernova stage?

The outcome of the star after the supernova stage will depend on the original mass of the star. In the case of an intermediate sized star, a very dense neutron star will be left behind. For the largest stars, those that are the most massive (more than 10-20 times the size of the sun), the gravitational collapse will be so complete that all that will be left will be a black hole. Ancient astronomers recorded these mysterious bright objects in the sky as divine omens.

Who was the first to record the appearance of a supernova?

The Danish astronomer, Tycho Brahe, in November 1572 noticed a new star in the constellation Cassiopeia. The star was so bright, it was even observed in the daytime.

It wasn't until the mid 1600s, using telescopes and newly developed star charts, that astronomers learned that these bright spots were not newly formed stars, but rather existing stars that had gained brightness. We now know that these super bright stars are the explosive end to a massive star.

When did we first understand the origins of supernovas?

The pioneering work of Fritz Zwicky and Walter Baade in the 1930s led to our understanding of supernovas and novas. They were able to measure the difference between novas and supernovas, and suggested that neutron stars were the remains of supernovas.

It was also concluded that the supernova was a rare event, happening only one to three times every thousand years per galaxy. Newer studies concluded that this event occurs more frequently, every fifty years per galaxy. The newer studies concluded that supernovas are less visible because of interference by interstellar clouds.

It was also in the 1930s that Indian born American astronomer Subrahmanyan Chandrasekhar put together the sequence of events leading up to the formation of supernova. He also formulated the method for calculating the mass that would determine the outcome of a star, whether it became a white dwarf or neutron star.

Concept Reinforcement:

1. Explain the concept of a supernova.

2. State how frequently supernovas occur.

3. Describe what happens to a star after the supernova stage.

Chapter 27 – Quasars

Chapter Objective:

- Understand and explore quasars

What is a quasar?

A quasar is a compact object that is beyond our galaxy. They are so distant that the light they emit takes several billion years to reach us. In fact they are so bright that they shine more brightly than one hundred galaxies together.

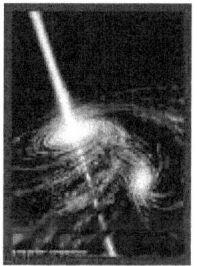

Images courtesy of NASA

Where does the word "quasar" come from?

Quasar is a form combined from quasi-stellar radio sources. Some quasars have indeed been observed through radio telescopes. Even though only about 10% of all quasars emit radio waves, they also emit energy in the forms of visible light, infrared and ultraviolet radiation, x-rays, and even gamma rays.

Quasars were first identified in the early 1960s by astronomer Alan Sandage, who photographed an area of the sky that had a very unusual spectrum and which was first believed to have been a star. That theory soon changed when the detectable wavelengths of the radiation being emitted were predominantly towards the red end of the visible light spectrum. This indicated the object was moving away from the point of origin and was apparently doing so with great speed.

The Dutch astronomer Maarten Schmidt observed a quasar through a 200 inch Hale telescope while at the California Institute of Technology. His observations and calculations placed the observed object at two billion light years away from Earth. It was emitting as much energy as one trillion suns and was about the size of our solar system.

What makes quasars of interest for study?

Quasars do give us a glimpse of the early universe. If we consider looking at an object a billion light years away, we can consider this like looking through a window back in time, since it took a billion years for the light to reach us. So when we see the light from a quasar, it is like looking at a relic from a period that followed the big bang.

How do astronomers explain quasars today?

Astronomers now believe that a quasar is formed from the collision of two distant galaxies. When this occurs, one galaxy then creates a black hole in the other with a mass of about one hundred million suns. Gas, dust, and stars are continually pulled into the black hole, where the temperature rises to hundreds of millions of degrees, and emits large amounts of radiation.

The brightest quasar to date, located in the constellation Draco, shines as bright as 1.5 quadrillion suns.

Concept Reinforcement:

1. Describe a quasar.

2. Explain how quasars were first identified.

3. Discuss why quasars are interesting to scientists.

Chapter 28 – Pulsars

Chapter Objective:

- Understand and explore pulsars

What is a pulsar?

A pulsar is a neutron star that emits beams of radiation that sweep through the Earth's line of sight. It is also an endpoint to stellar evolution. It is the spinning of a neutron star and its increasing magnetic field, which causes the star to act as a giant magnet. It will emit radiation out of its magnetic poles, and if the magnetic axis is tilted in a certain way, the rotation of the star's signal is visible from Earth.

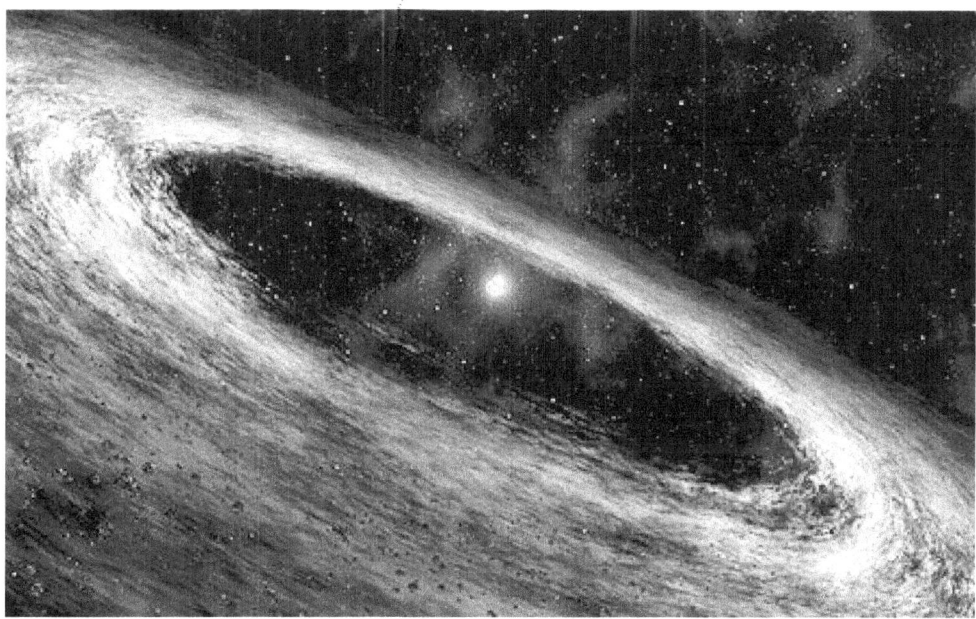

An artist's concept of a planet-forming disk around pulsar. Image courtesy of NASA

How were pulsars discovered?

In the mid 1960s, a Cambridge University PhD student Jocelyn Susan Bell Burnell, and her supervisor Antony Hewish, built a radio telescope that was designed to track quasars. They were able to track and record changing energy signals on long rolls of paper. It was observed that some of the signals would pulsate regularly, where others had been consistent and continuous. The objects were named pulsars, and they theorized that the pulsars might be dwarf or neutron stars. It was at the end of the following year that two astronomers Thomas Gold and Franco Pacini concluded that neutron stars were the source of these signals that Bell Burnell and Hewish had detected.

It was concluded that neutron stars must be the source of pulsars because neutrons are very dense and rotate very quickly. As the spinning intensifies, the magnetic pull increases and emits radiation that is visible from Earth, if the magnetic axis is tilted a certain way.

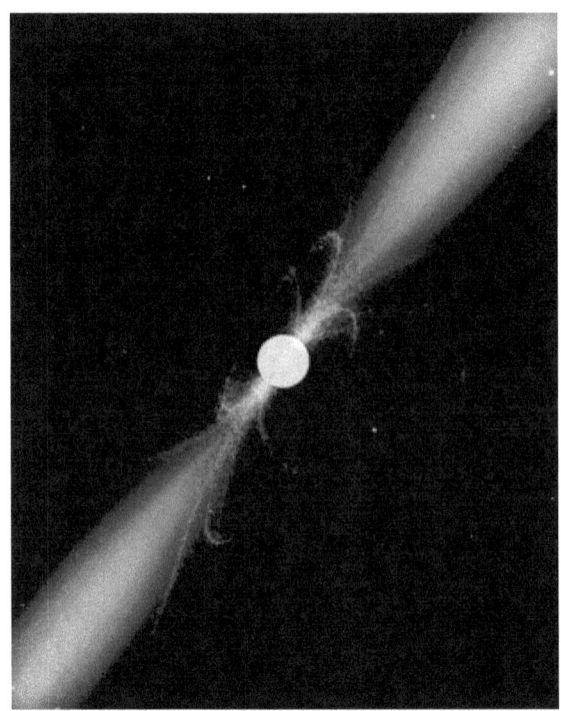
Pulsar image courtesy of NASA

Pulsars tend to have large magnetic fields and spin rapidly. It is the loss of the pulsar's spin energy, which eventually appears as radiation across the electromagnetic spectrum, which includes gamma rays.

Pulsars will perhaps be better understood once the emission of gamma ray energy can be better detected and monitored in the light curve, by creating models of how this light is created.

More than 1500 pulsars have been catalogued, some in areas where there were once supernovas. Scientists now believe that our galaxy alone may contain more than one hundred thousand active pulsars.

There are differences in the brightness of pulsars based on their emissions. All pulsars are neutron stars, but X-ray pulsars differ in the in their emissions and the level of their intensity.

Concept Reinforcement:

1. Describe a pulsar.

2. Explain how pulsars were discovered.

3. Explain why neutrons are considered the source of pulsars.

Chapter 29 – Black Holes

Chapter Objective:

- Understand and explore black holes

What are black holes?

Artist's rendering of a black hole image courtesy of NASA

Black holes are perhaps one of the most interesting phenomena in the cosmos. It is a place where space and time meet and stretches out for infinity. Black holes are not visible, yet they account for about 90% of the whole universe.

A black hole is all that remains of a massive star that has used up its nuclear fuel and collapsed under gravitational pull into a single point. When anything gets too close to a black hole, it gets pulled in, and gets stretched to infinity and will remain forever trapped. Even light cannot escape a black hole.

How does a black hole form?

A star must first be at least two or three times the mass of our sun. A star collapses once the nuclear fuel is exhausted. The nuclear fusion process forces the gravity outward from the star's core. A star that is of average mass, like our sun will end as a white dwarf. A star that is five to eight times the mass of the sun will explode to produce a supernova, shedding its mass and becoming a neutron star.

A star that is ten to forty times the mass of the sun will produce a gravitational collapse that is so complete, the black hole is the only outcome after the supernova.

As a giant star collapses, its mass is so concentrated, with the force of gravity incredibly strong. The collapsed star's surface is known as the event horizon, and anything that crosses that threshold can not escape the gravitational pull being exerted upon it. Event horizons are the point of no return for any particle. Once the particle crosses the event horizon into the black hole, it will never cross back. An interesting feature of the event horizon is that, while objects may still be orbiting the black hole within the event horizon, they are not visible from outside the event horizon. It is also not possible to know when an object will cross the event horizon because it is not visible.

Black holes can continue to grow after they are formed. This occurs as interstellar dust and cosmic background radiation are pulled into the black hole. These additions result in minimal growth of the black hole. If, however, the black hole merges with other large objects, such as stars, the size of the black hole will increase substantially.

Concept Reinforcement:

1. Describe a black hole, including how much of the Universe they comprise.

2. Explain how a black hole forms.

3. Discuss how black holes grow and affect the area surrounding them.

Chapter 30 – Wormholes

Chapter Objective:

- Understand and explore wormholes

What are wormholes?

Once thought to be more science fiction than science fact, wormholes were first explained in 1935 by physicists. Albert Einstein and Nathan Rosen understood that general relativity allows for the existence of "bridges," or wormholes, or space-time tubes, which actually act as shortcuts connecting distant regions of space-time. These "bridges" could exist in theory "intra" or "inter" universe.

Albert Einstein

So by definition, you could travel between the two regions faster than a beam of light would be able to if it moved through normal space and time.

Thus, wormholes became the perfect subject for time travel and the material that science fiction writers could use in their stories.

Artists Rendering of a Wormhole

It was only recently that the theory that wormholes only existed for a short period of time was disputed. The possibility does exist that wormholes could indeed be utilized for time and space travel.

A wormhole is still considered a hypothetical tunnel, which connect two different points in spacetime in such a way that a trip through a wormhole would take less time to complete than a journey between the same starting and ending points in normal space.

Wormholes are very unstable and would probably collapse instantly if the tiniest amount of matter attempted to pass through. A possible solution to this problem would be the use of exotic matter (a hypothetical type of matter that has both a negative energy density and a negative pressure that exceeds the density of the energy) to prevent the wormhole from collapsing on itself.

It is theorized that wormholes may exist today, and that they were formed in the Big Bang, spanning small or vast distances in space.

Concept Reinforcement:

1. Describe the first theory of wormholes.

2. State how the theory of wormholes has changed.

3. Explain why wormholes are problematic for spacetime travel, even though it is a topic of interest to scientists and science fiction writers.

Chapter 31 – The Universe

Chapter Objective:

- Introduce the basic principles of astronomy in relation to the universe

When we talk about the study of the universe, we are referring to the field of cosmology. Cosmology is the study of the origin, evolution, and structure of the universe. It is a science in and of itself that grew out of simple observations. It evolved along the way with mathematical theories, technological advances, and space exploration.

The study of the universe includes many concepts, including the Big Bang Theory, galaxies, energy, and the theory of relativity.

Image courtesy of NASA

How long have humans studied cosmology?

The earliest theories of our universe were developed by early astronomers over a period of about 3,500 years–from 2200 B.C. to 1200 A.D. The astronomers were in Babylon, China, Greece, Italy, India, and Egypt. These astronomers made observations without the aid of sophisticated equipment, typically with just the naked eye or very simple magnifying devices.

The universe is everything that physically exists. In other words, it is the entirety of space and time, all forms of matter, energy and momentum, the physical laws and the constants that govern them.

Image courtesy of NASA

How old is the universe?

The age of the Universe has been a subject of religious, mythological and scientific importance. On the scientific side, Sir Isaac Newton's guess for the age of the Universe was only a few thousand years. Einstein, the developer of the General Theory of Relativity, preferred to believe that the Universe was ageless and eternal. However, in 1929, observational evidence proved his fantasy was not to be fulfilled by Nature.

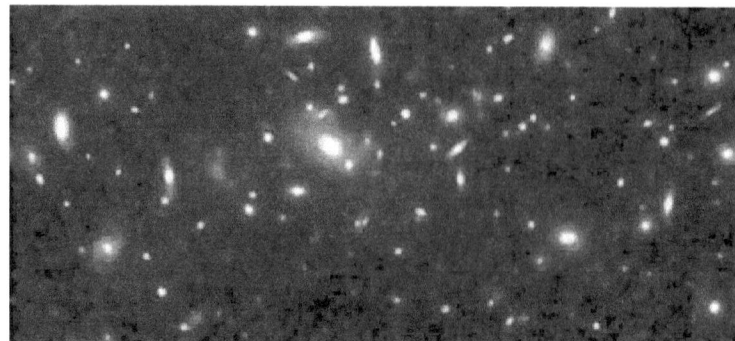

Image courtesy of NASA

Astronomical observations predict that the universe is at least 13 billion years old and about 93 billion light years across. The most discussed theoretical event that started the universe is called the Big Bang. The Big Bang theory was first elaborated on by Belgian astronomer and Jesuit priest Georges Henri Lemaitre in the late 1920s. Prior to the big bang, all matter and energy was concentrated in one point of infinite density.

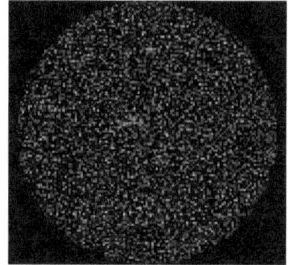

Image courtesy of NASA

Development of theories about the structure of our universe is an important topic for discussion, as are the concepts related to the idea that the big bang might happen yet again.

The most fundamental physics concept that is applied to the study of the Universe, is Albert Einstein's Theory of General Relativity. This theory explains the structure and formation of the universe.

Concept Reinforcement:

1. Describe the field of cosmology.

2. Explain the big bang theory.

3. Discuss the importance of Einstein's Theory of General Relativity.

Chapter 32 – The Structure of the Universe

Chapter Objective:

- Apply astronomy concepts to explain the structure of the universe

In the study of cosmology, the term large-scale structure refers to the characterization of distribution of matter and light that is observable on the largest scale. The size of the universe has often been a topic for speculation. In the 1990s, astronomers discovered that the universe is larger than they thought. By making maps of the universe, it was discovered that great "sheets" of galaxies in clusters and super-clusters fill areas hundreds of millions of light years in diameter. These sheets of galaxies are separated by large empty spaces of darkness, up to four hundred million light years across.

Artist's Rendering of Sunrise from Space

If we think of the universe in very simple terms, for example, a very large ball, it is possible to observe the ball in three dimensions of space. The outer surface of the ball has the geometry of a sphere in two dimensions, because there are only two independent directions of motion along the surface (from pole to pole and around the largest diameter of the sphere). If measurements of the universe were made, it would be clear that you were measuring the curved surface of a large sphere.

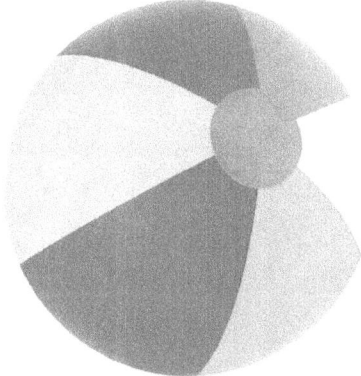

The Universe can be thought of as a large three dimensional ball.

It is this idea of the curvature of the surface of the ball that can be applied to the whole universe.

This was the main inspiration behind Albert Einstein's theory of general relativity. One concept is the idea of space and time being unified into a single geometric entity called space-time. Stated simply, space-time describes all of space and all of time. This includes everything that has ever happened, and everything that ever will happen.

When cosmologists solve the Einstein equation for the space-time geometry of our Universe, they consider three basic types of energy that could curve space-time: vacuum energy, radiation, and matter.

Vacuum energy is the background energy of space, whether or not there is matter present. Radiation is the energy emitted by stars and other space matter. Matter is the substance that makes up the parts of the universe we can see and touch, such as planets, stars, comets, moons, asteroids, etc.

The theory of general relativity explains many previously unexplained phenomena, such as the orbits of certain planets, pulsars, the expansion of the universe, time and the ability of gravity to bend rays of light.

Concept Reinforcement:

1. Explain what scientists learned in the 1990s about the size of the universe.

2. Define space-time.

3. List the three types of energy that can curve space-time.

Chapter 33 – The Formation of the Universe

Chapter Objective:

- Apply astronomy concepts to explain the formation of the universe

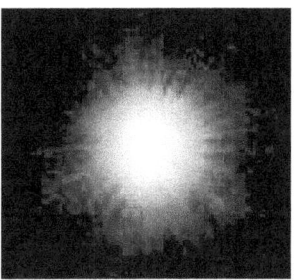

How was the Universe created?

This question has been asked by astronomers for as long as they have observed the skies. As far as we can tell, the expansion of the Universe started many billions of years ago from a very hot, very small state. From that hot, small state, it mushroomed and evolved into the Universe we know today. Cosmologists call that process of expansion the **Big Bang** because at some phases, especially in the beginning, the process was rather like an explosion. Much of understanding of the Big Bang is the result of extrapolation of our existing knowledge of particle physics today, and projections from the mathematical model of an expanding universe based in Einstein's Theory of General Relativity.

The two most popular theories about the formation of the universe are the big bang theory and the steady state theory.

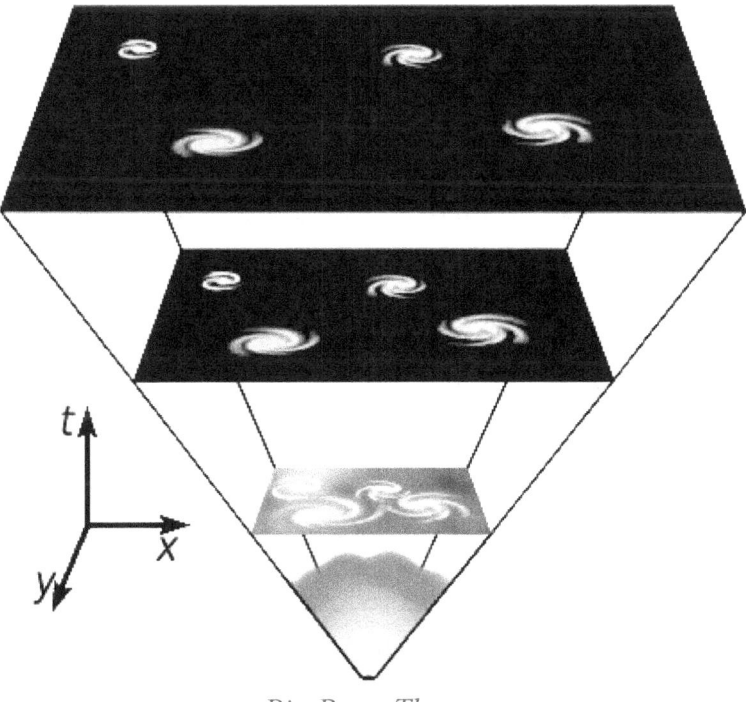

Big Bang Theory

The big bang theory is the most widely accepted creation theory, which suggests that the universe came into being with a big explosion fifteen to twenty billion years ago. Gravity came into being almost immediately, followed by atoms, stars, and galaxies. According to the big bang theory, the solar system formed 4.5 billion years ago from a cloud of dust and gas.

The other theory is known as the steady-state theory, which claims that all matter in the universe has been created continuously, a little at a time, at a constant rate, from the beginning of time.

Georges Henri Lemaitre

The steady-state theory was first proposed in the late 1920s by Belgian astronomer and Jesuit priest Georges Henri Lemaitre. The theory has also stated the universe is structurally the same all over, and has been consistent throughout time. Thus, the steady state theory assumes the universe is infinite, unchanging, and will last forever.

The steady-state theory was further developed by Thomas Gold and Hermann Bondi in 1948 when they asserted that even through the universe is expanding, it does not change in appearance over time. This is called the perfect cosmological principle, which infers that the universe has no beginning or end. In 1963, the discovery of quasars, very distant, bright, star-like objects, disproved the steady-state theory because it did not predict the existence of quasars. The big bang theory predicted the presence of quasars.

Concept Reinforcement:

1. List the two primary theories about the formation of the universe.

2. State which theory is the most accepted and why.

3. What impact did the discovery of quasars have on the steady state theory?

Chapter 34 – The Expanding Universe

Chapter Objective:

- Apply astronomy concepts to explain the expanding universe

What do we mean when we say "expanding universe?"

For thousands of years astronomers asked the basic questions about the size and age of the universe. Does the universe go on forever, or does it have an edge somewhere? Has the universe always existed, or did it come to being some time in the past?

In 1929, Edwin Hubble, an astronomer at Caltech, made a discovery that led to answers for these questions based on scientific fact. He discovered that the universe is expanding.

The ancient Greeks had recognized that it was difficult to imagine what an infinite universe might look like. However, if the universe is finite and you stick your hand out to the edge, where will your hand go? Both alternatives presented problems to the ancient Greeks.

The discovery of the expanding universe theory came after physicists and mathematicians working on Einstein's theory of gravity discovered the equations had solutions that described an expanding universe. Redshift is a key concept in modern astronomy. It is a result of the increases in the wavelength of light as it crosses the universe, which is continually expanding, causing the light to move to the lower energy (red) part of the light spectrum. In the solution to the equation, the light coming from distant objects will be redshifted as it travels through the expanding universe. The redshift will increase with increasing distance to the object.

In 1929 Edwin Hubble measured the redshifts of a number of distant galaxies. He also measured their relative distances by measuring the apparent brightness of a class of variable stars called Cepheids in each galaxy. Hubble measured the redshift as a plot against relative distance. The finding was the redshift of distant galaxies increased as linear function of their distance. The only explanation for this observation is that the universe is expanding.

Once scientists realized that the universe was expanding, they realized that the universe had to have been smaller in the past. Taking this to the extreme, at some point in the past, the entire universe would have been a single point. This point, later called the big bang, was the beginning of the universe as we know it today.

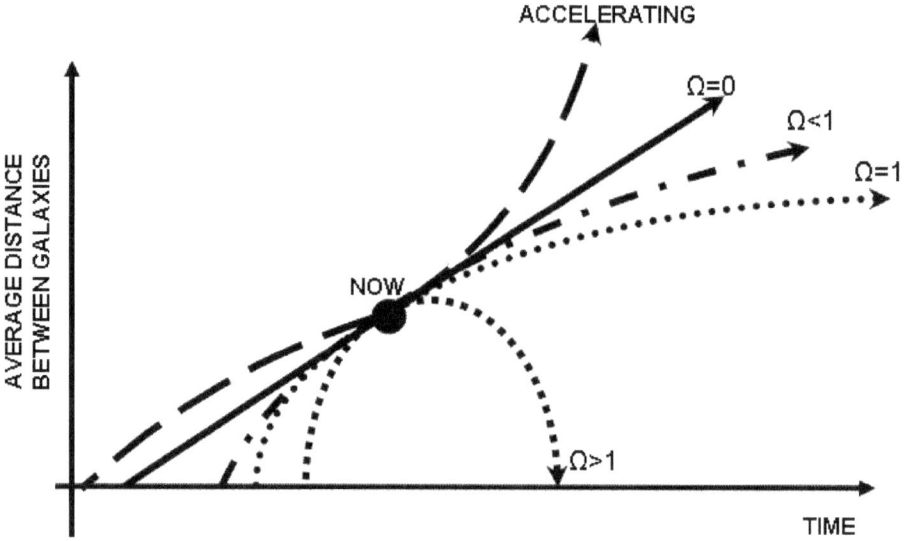

The expanding universe is finite in both space and time. The universe is in a constant state of change. The expanding universe is a new idea based on modern physics that brought new light to the concept of the infinite universe.

What are the properties of the expanding universe?

The equations that describe the expanding universe have three possible solutions and the fate that ultimately befalls the universe can be determined by measuring how fast the universe expands relative to how much matter the universe contains.

The three possible types of expanding universes are called open, flat, and closed universes.

In an open universe expansion goes on forever.

In a flat universe, expansion also continues, but the expansion rate would slow to zero after an infinite amount of time.

In a closed system, the universe would eventually stop expanding, and recollapse on itself, potentially leading to another big bang.

In all of the three cases, the expansion slows, and gravity is the force that causes the slowing.

Concept Reinforcement:

1. What did Edwin Hubble discover in 1929?

2. Describe redshift.

3. List the three possible types of expanding universes and their impact on the universe.

Chapter 35 – The Big Bang Theory

Chapter Objective:

- Apply astronomy concepts to explain the Big Bang Theory

The Big Bang Theory is the dominant scientific theory about the origin of the universe. According to the theory, the universe was formed 10 to 20 billion years ago from a cosmic explosion that hurled matter in all directions.

Much of understanding the Big Bang is extrapolating between knowledge of particle physics today and projections from the mathematical model of an expanding universe described by the theory of general relativity. Albert Einstein developed the mathematical model used to describe how fast the Universe will expand, including size and time, given the energy density of matter and radiation at that time. We base our guesses about the matter and radiation density of the early Universe based on the ancient light reaching us from the past in our night skies, and what we have learned about elementary particle physics, through theory and experiment.

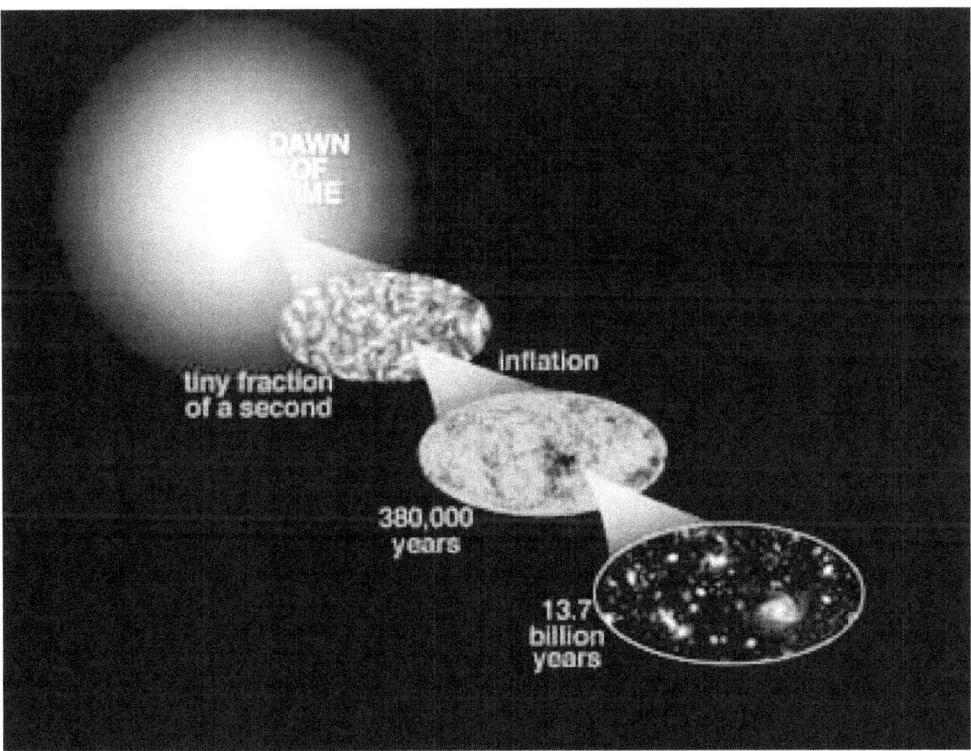

The big bang theory image courtesy of NASA

This theory, first proposed by the Belgian priest Georges Lemaitre in 1927, was the first to propose that the universe began with the explosion of a primeval atom. This proposal came after observing a red shift in distant nebulas by astronomers to a model of the universe based on relativity.

The big bang was initially suggested because it explains why distant galaxies are traveling away from us at great speeds. The theory also predicts the existence of cosmic background radiation (which is the glow left over from the explosion itself), as well as pulsars.

The Big Bang Theory is widely accepted, however, it may never be proved.

Concept Reinforcement:

1. Explain the Big Bang Theory.

2. Describe how Einstein's Theory or Relativity supports the Big Bang Theory.

3. List three things the Big Bang Theory explains about the universe.

Chapter 36 – Dark Matter

Chapter Objective:

- Apply astronomy concepts to explain dark matter

It is believed that most of the matter in the universe is dark. Dark matter is matter that does not emit light because it does not interact with the electromagnetic force, however it does have gravitational effects on visible matter, such as planets.

If dark matter can't be seen, why do we think it exists at all?

Its presence is inferred indirectly from the motions of astronomical objects, stellar, galactic, and galaxy clusters and superclusters. Many of the observations of galaxies and structures larger than galaxies lead to the conclusion that there is more matter than can be observed. Dark matter is important in understanding the rotational speeds of galaxies, orbital velocities of clusters of galaxies, and other astronomical phenomena.

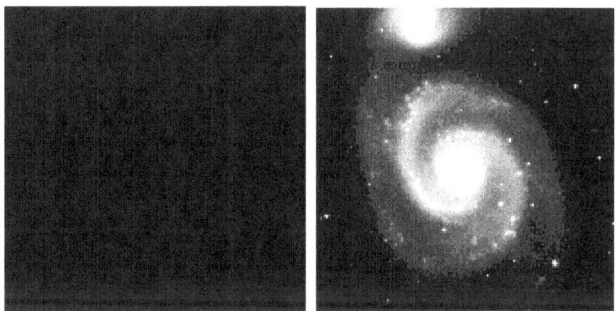

Images courtesy of NASA

Dark matter also has important consequences for the evolution of the Universe and the structure within it. The total mass and energy in the universe affects the possible outcome or fate of the galaxy based on the three possible types of expanding universes: open, flat and closed.

Dark matter candidates are usually split into two broad categories: baryonic and non-baryonic.

Baryonic matter is considered to be ordinary matter, which is made of baryons (protons and neutrons) that do not emit detectable radiation. Some examples include non-luminous gas and brown dwarfs.

Non-Baryonic matter is further subdivided into hot dark matter and cold dark matter.

Hot dark matter is made up of lightweight particles that move near the speed of light. An example is thought to be the neutrino.

Cold dark matter, on the other hand, is made of extremely massive particles that move very slowly. Physicists have dubbed these particles Weakly Interacting Massive Particles, or WIMPs. Three types of cold dark matter have been described, although never observed. These include photinos, which have 10-100 times the mass of a proton; axions, carriers of force that have mass; and quark nuggets, which are non-baryonic collections of quarks.

Concept Reinforcement:

1. Describe why physicists think dark matter exists.

2. Explain baryonic dark matter and its key characteristic.

3. Discuss the difference between non-baryonic cold and dark matter.

Chapter 37 – Background Radiation

Chapter Objective:

- Apply astronomy concepts to explain background radiation

What is background radiation?

Background radiation is the ionizing radiation that exists all around us. There are both natural and artificial sources for this radiation. The light from the sun is a form of radiation. Ionizing radiation is made from high-energy particles that are able to ionize (remove) at least one electron from an atom, giving the atom or molecule an electrical charge (positive or negative).

Ionizing radiation is emitted from a variety of natural and artificial sources, and these include:

1. The Earth: food, water, building materials, plants, etc.

2. Space–cosmic rays

3. Sources in the atmosphere–radon gas from the Earth's crust

There are many sources of background radiation that are man made. About 15 % of the background radiation the general public experiences results from clinical application of medical X-rays and nuclear medicine.

Another 3% comes from other man-made sources, including smoke detectors, self-luminous signs, radioactive contamination, nuclear power plant accidents, normal operations of a nuclear power plant, emissions from nuclear medicine, and emissions from the improper radioactive waste disposal.

What is natural background radiation?

Natural background radiation comes from two main sources: cosmic radiation and terrestrial sources. The worldwide average background dose for humans is estimated at about 2.4 millisievert (mSv) per year. A sievert is the international standard measure of a dose equivalent. A millisiervert is $1/1000^{th}$ of a sievert. This amount is still far higher than the exposures people receive from man-made radiation sources, which is equivalent to 0.01 mSv per year.

The levels of background radiation will vary depending on your location. For example, in Ramsar, Iran, the dosage has been reported to be about 260mSv per year.

Background radiation may simply be any radiation that is pervasive. A particular example of this is the cosmic microwave background radiation, which is a nearly uniform glow that fills the sky in the microwave part of the spectrum; stars, galaxies and other objects of interest in radio astronomy stand out against this background.

In a laboratory, background radiation refers to a measured value from any source that can affect an instrument when a radiation source sample is not being measured. This background rate, which must be established as a stable value by multiple measurements, usually before and after sample measurement, is subtracted from the rate measured when the sample is being measured.

Background radiation is used to measure radiation exposure for workers whose jobs involve some exposure to radiation. The actual occupational doses for workers are measured by subtracting the background radiation from the total radiation resulting from their jobs. This includes both "natural background radiation" and any medical radiation doses. This value is not typically measured or known from surveys, such that variations in the total dose to individual workers are not known. This can be a significant factor in assessing radiation exposure effects in a population of workers who may have significantly different natural background and medical radiation doses. This is most significant when the occupational doses are very low.

Concept Reinforcement:

1. Describe background radiation.

2. List both natural and manmade sources of background radiation.

3. Explain how radiation is measured in a laboratory.

Chapter 38 – The Doppler Effect

Chapter Objective:

- Apply astronomy concepts to explain the Doppler Effect

The Doppler Effect can be described as a change in pitch of sound waves, resulting from a shift in the frequency of the sound waves. The Doppler Effect is named after the Austrian mathematician and physicist, Christian Doppler.

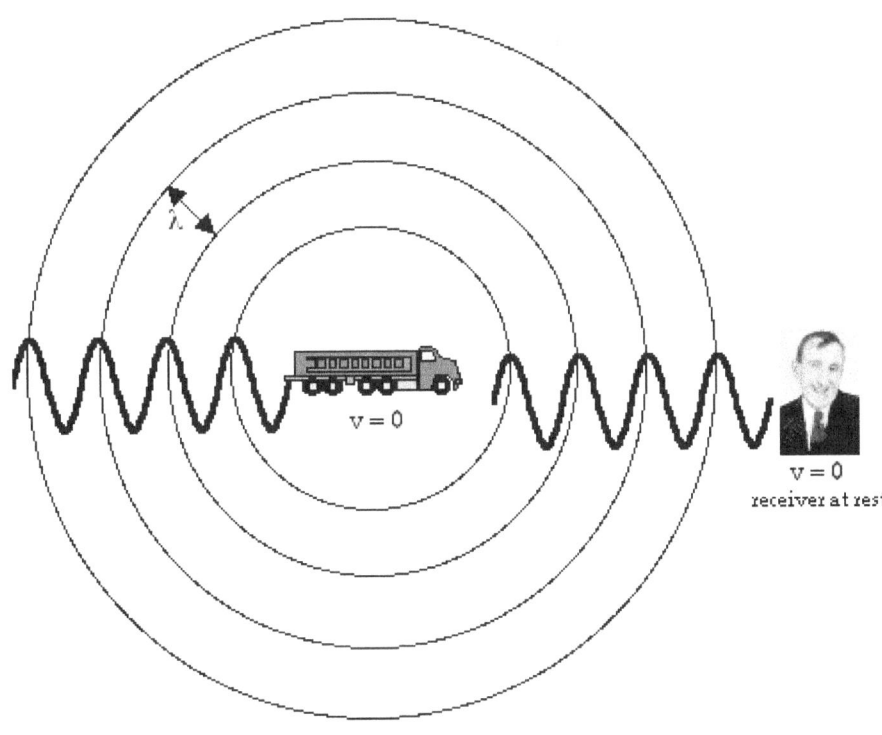

The Doppler Effect

The Doppler Effect can also be evaluated in astronomy, as electromagnetic radiation is measured from the moving object. As the radiation emitted by an object moving toward the observer is squeezed, its frequency appears to increase and thus is said to be blueshifted. In contrast, the radiation emitted by an object moving away is stretched or said to be redshifted. Blueshifts and redshifts exhibited by stars, galaxies and gas clouds also indicate their motions with respect to the observer.

In astronomy, the Doppler Effect was originally studied in the visible part of the electromagnetic spectrum. Radiation is blueshifted when its wavelength decreases and redshifted when its wavelength increases.

Astronomers use Doppler shifts to calculate precisely how fast stars and other astronomical objects move toward or away from Earth.

Shifts in frequency result from other factors besides relative motion; one associated with strong gravitational fields and known as Gravitational Redshift, and the other known as Cosmological Redshift, which results from the expansion of space following the Big Bang.

Concept Reinforcement:

1. Describe the Doppler Effect.

2. Explain how the Doppler Effect relates to red and blue shifts.

3. Discuss the other forces that can cause redshifts.

Chapter 39 – Galaxy Clusters

Chapter Objective:

- Apply astronomy concepts to explain galaxy clusters

Galaxies are usually found in groups, although there are some that exist on their own. They are also found in larger agglomerations known as clusters. Clusters are held together by gravity. The Local Group consists of our galaxy, Andromeda (a larger spiral galaxy), and several smaller satellites, which include the Large and Small Magellenic Clouds.

Image courtesy of NASA

Regular clusters have a couple of key features: a concentrated center core and a defined spherical structure. These are further subdivided according to the number of galaxies and how dense they are in relation to the proximity to their center (known as the Abell radius). The mass of a regular cluster is estimated at about the same as a million billion suns.

Image courtesy of NASA

Irregular clusters have no well-defined center. They have a similar range of sizes as the regular clusters, but they typically have less mass (1,000 to 100,000 million suns).

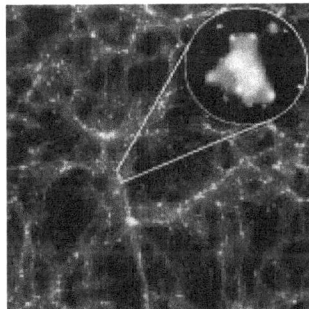

Image courtesy of NASA

Superclusters are chains of clusters (usually about 12). Superclusters have a mass equivalent to that of about 10 million billion suns. The superclusters then form a network that contains about 90% of the galaxies in the universe.

Voids, sheets and filaments define the structures of the galaxies. Redshift surveys have shown the universe to have a very bubbly structure. This structure consists of sheets and filaments, which contain the galaxies, with voids as the dominant feature.

Image courtesy of NASA

The Hubble Space Telescope has been utilized to take very deep galaxy surveys. The image above shows galaxies just a couple of billions years after the Big Bang. The Hubble Space Telescope will continue to help us learn about our universe as it moves through space.

Concept Reinforcement:

1. Describe how a cluster of galaxies stays together.

2. Define a supercluster and how superclusters interact.

3. List the primary components that form a galaxy.

Chapter 40 – Voids in Space

Chapter Objective:

- Apply astronomy concepts to explain voids in space

Voids are the areas in space that emit few or no electromagnetic signals. They are common in all galaxies and comprise most of the free space in the universe. Recently scientists have found the largest void in space, or largest hole in the universe. The void is nearly a billion light years across, and does not contain normal and dark matter. This finding does challenge the theories of large-scale structure formation in the universe.

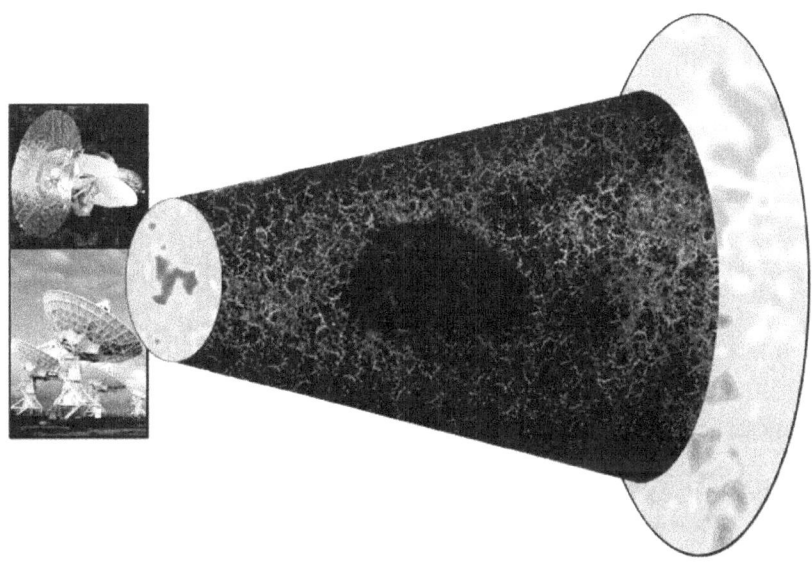

Huge Void Implicated in Distant Universe Image courtesy of NASA

The discovery was made by accident by Lawrence Rudnick and colleagues of the University of Minnesota in Minneapolis, USA. Rudnick and colleagues discovered a "cold spot", which turned out to be an unexplained anomaly in the map of the cosmic microwave background (CMB), which was created by NASA's WMAP satellite. Rudnick's group began looking for radio sources such as radio galaxies and quasars in the direction of the cold spot. Radio sources track the distribution of mass in the universe. They are considered the markers for galaxies, clusters of galaxies and dark matter.

This void is about 6 to 10 billion light years away, and is larger than others found previously. There were few or no radio sources in a volume equivalent to a billion light years in diameter. The lack of radio sources means that there are no galaxies or clusters in that volume, and since the CMB is cold also suggests that the area lacks dark matter. This particular void makes it the largest one to date at 40 times larger than the previous record.

What makes this latest void, discovered by Rudnick so interesting?

The CMB is leftover radiation from the big bang, and some cosmologists have said that the cold spot is a problem for the theories of the early universe. The void could have been produced billions of years after the big bang. Rudnick proposes to look at this phenomenon more as a problem in the time of structure formation, rather than a problem left over from the early universe.

What is the future of the universe?

Will the universe expand forever? Or will the cosmos eventually collapse into another Big Bang? The answer depends on whether the universe contains enough mass—and thus gravity—to slow and reverse the expansion. Unfortunately, astronomers can't directly measure the total mass of the universe because most of it appears to be in the form of dark matter, which does not emit light and is therefore not measurable. Barbara Ryden, an astronomer at Ohio State University is currently doing a sky survey. By the decade's end, when she has completed the data collection for the sky survey, she should be able to foretell the future of the universe.

Concept Reinforcement:

1. Describe a key characteristic of a void.

2. Explain why the void discovered by Rudnick is so interesting.

3. What does Dr. Ryden hope to accomplish with her sky survey?

Chapter 41 – Energy in the Universe

Chapter Objective:

- Apply astronomy concepts to explain energy in the universe

What types of energy exist?

Astronomers often describe the universe as expanding, and also state that energy changes form from one form to another. Mass is a form of energy. The mass of the universe is constant, thus the energy of the universe is constant, as well. The energy present at the beginning of the universe is still in the universe today.

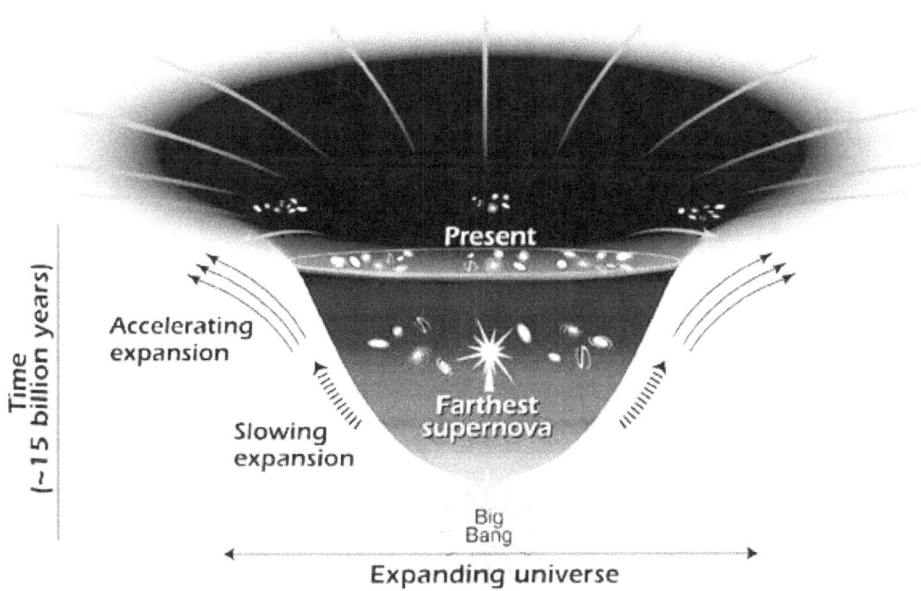

The diagram above shows the changes in the rate of expansion since the universe's birth 15 billion years ago. The more shallow the curve, the faster the rate of expansion. The curve changes noticeably about 7.5 billion years ago, when objects in the universe began flying apart at a faster rate. Astronomers theorize that the faster expansion rate is due to a mysterious, dark force that is pulling galaxies apart. Image courtesy of NASA/STScI/Ann Feild.

More recently, dark energy has been discussed as being one of the predominant forms of energy in the galaxy. Dark energy is the mysterious energy in the universe with unusual anti-gravitational properties and is presumed to have a large role in the expansion of the universe. Dark matter is estimated to account for more than 70% of the mass energy of the universe. These findings are based on calculations concerning the age of the oldest stars, with expansion measurements and the overall geometry of the universe.

How do you conclude that such matter exists, when you can't physically observe it? Observations of distant supernova have also suggested that dark energy dominates the universe.

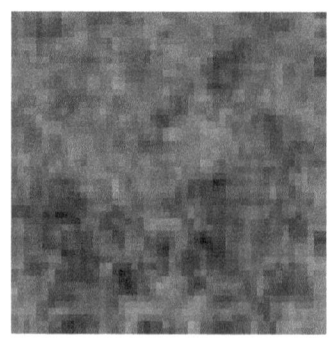

Image courtesy of NASA

These and other observations led researchers to the conclusion that the universe is expanding more quickly than it has in the past. Astronomers Chaboyer and Krauss, proposed that the only explanation for an accelerating universe is that the energy content of a vacuum is non-zero with negative pressure, or dark energy.

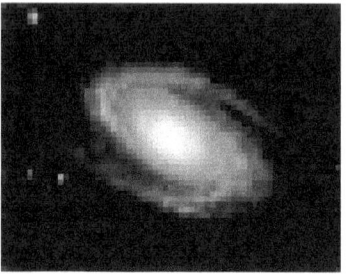

Image courtesy of NASA

Concept Reinforcement:

1. Define dark energy.

2. State the percentage of the mass energy of the universe that is dark energy.

3. Discuss the role that dark energy plays in the expansion of the universe.

Chapter 42 – Einstein's Ideas

Chapter Objective:

- Explore Einstein's ideas in astronomy

Albert Einstein 1905

Who is Albert Einstein?

Albert Einstein was a physicist who developed the Theory of Relativity. The theory applies where motion is constant, meaning there is no speeding up or slowing down. According to this theory, time and space are not fixed. Einstein explained that four dimensions are required to describe the universe; three dimensions in space plus time.

In November of 1919, at the age of 40, Albert Einstein's theory of relativity was confirmed, thanks to a solar eclipse. An experiment had confirmed that light rays from distant stars were deflected by the gravity of the sun in just the amount he had predicted in his theory of gravity, general relativity. General relativity was the first major new theory of gravity since Isaac Newton's more than 250 years earlier.

Einstein's theory of relativity (also known as general relativity) provided a new way of looking at the universe, in fact it was so revolutionary that Einstein himself had trouble accepting it.

What is Einstein's theory of general relativity?

Einstein's theory of time and space, special relativity, proposed that distance and time are not absolute. Published in 1915, general relativity proposed that gravity, as well as motion, can affect the intervals of time and of space. The key idea of general relativity, called the equivalence principle, is that gravity pulling in one direction is completely equivalent to acceleration in the opposite direction. For example, a car accelerating forwards feels just like sideways gravity pushing you back against your seat. An elevator accelerating upwards feels just like gravity pushing you into the floor.

Many of the predictions of general relativity, such as the bending of starlight by gravity and a tiny shift in the orbit of the planet Mercury, have been quantitatively confirmed by experiment. Two of the strangest predictions, impossible ever to completely confirm, are the existence of black holes and the effect of gravity on the universe as a whole (cosmology).

General relativity may be one the biggest leaps of the scientific imagination in history. General relativity had little foundation upon the theories or experiments of the time. No one except Einstein was thinking of gravity as equivalent to acceleration, as a geometrical phenomenon, as a bending of time and space. Many physicists believe that without Einstein, it could have been several decades or more before another physicist worked out the concepts and mathematics of general relativity.

Concept Reinforcement:

1. Explain Einstein's Theory or Relativity.

2. Describe how Einstein's theory was confirmed.

3. What was Einstein's unique perception of gravity?

Chapter 43 – Special Relativity

Chapter Objective:

- Apply astronomy concepts to explain special relativity

What is the Special Theory of Relativity?

Special relativity (SR- also known as the special theory of relativity-STR) is the physical theory of measurement with a frame or reference. The theory is based on objects in motion, relative to a fixed point, traveling at a significant fraction of the speed of light.

Einstein's theory of special relativity results from two hypotheses — the two basic postulates of special relativity:

1. The speed of light is the same for all observers, no matter what their relative speeds.

2. The laws of physics are the same in any inertial (that is, non-accelerated) frame of reference. This means that the laws of physics observed by a hypothetical observer traveling with a relativistic particle must be the same as those observed by an observer who is stationary in the laboratory.

Given these two statements, Einstein showed how definitions of momentum and energy must be refined and how quantities such as length and time must change from one observer to another in order to get consistent results for physical quantities such as particle half-life. To decide whether Einstein's postulates are a correct theory of nature, physicists tested whether the predictions of Einstein's theory match observations. Indeed many such tests have been made — and the answers Einstein gave are right every time!

The first postulate — the speed of light will be seen to be the same relative to any observer, independent of the motion of the observer — is the crucial idea that led Einstein to formulate his theory. It means we can define a quantity c, the speed of light, which is a fundamental constant of nature.

This second postulate is a basic assumption in all of science — the idea that we can formulate rules of nature that do not depend on our particular observing situation, or frame of reference. This does not mean that things behave in the same way on the earth and in space, e.g. an observer at the surface of the earth is affected by the earth's gravity, but it does mean that the effect of a force on an object is the same independent of what causes the force and also of where the object is or what its speed is.

Einstein's theory is now very well established as the correct description of motion of relativistic objects, that is, those traveling at a significant fraction of the speed of light.

Because most of us have little experience with objects moving at speeds near the speed of light, Einstein's predictions may seem strange. However, many years of high energy physics experiments have thoroughly tested Einstein's theory and shown that it fits all results to date.

Concept Reinforcement:

1. State the theory of special relativity.

2. State the two basic postulates of special relativity.

3. Explain the two postulates of special relativity.

Chapter 44 – Theories of Time Travel

Chapter Objective:

- Apply astronomy concepts to explain the theories of time travel

H.G. Wells

The Time Travel Concept

The concept of time travel is a very interesting, yet a complicated topic. Time travel is the concept of moving between two points of time, either forward or backward, without experiencing the intervening time. Many of us are familiar with the many ideas of time travel that have been presented in science fiction novels. In fact one of the most well known stories about time travel is a novel by the late novelist H.G. Wells first published in 1895. H. G. Wells introduced the concept of a vehicle that allows an operator to travel purposely and selectively through time.

The question everyone wants answered is this: How do you travel in time? It is a simple question, but the answer is not quite as simple.

There are so many aspects to this particular question. Let's begin with how time is calculated.

Artists rendering of a Time Travel Device

Logic would suggest that if time travel were ever to exist, then it already does. If time travel is possible, then the methods required to do so will eventually be devised. Maybe it will take another 10 thousand years to discover the secrets of time travel, but if it's possible, then it's inevitable. Which suggests that time travelers are already visiting us... and visiting our past. What does that mean? To comprehend this, we must look at the many possible repercussions of time travel.

What follows are basic theoretical concepts of time travel:

Theory A–**Fate (Circular Causation)**–Travel back in time to save someone's life only to discover that it cannot be avoided, or worse yet, you were in fact the cause of the person's death in the first place. This is amongst the most plausible theories.

Theory B–**Alternate Universe**–Travel back in time to save someone's life, succeed, return to your time to discover that nothing has changed... you've only changed the timeline of an alternate quantum reality. This theory is also amongst the most plausible.

Theory C–**Success**–Travel back in time to kill your great-grandfather and succeed. This theory is very unlikely since if you were to successfully kill your great-grandfather, you would inevitably never be born, and therefore never go back in time to kill your great-grandfather. Thus the paradox and the implausibility of time travel.

Theory D–**Observer Effect**–Travel back in time to alter history and succeed, but the only persons capable of differentiating between the reality left behind and the new reality are those directly associated with the time travel... the time traveler. The extent of the paradox rests in how the time traveler is affected. Existing "out of time", he may not be affected by whatever changes he inflicts on the timeline, thus the time traveler himself becomes a stranger in this "new" present. He may, in fact, go back in time, kill his great-grandfather, and return to the present to discover that there are no records of his own existence.

The theories of general and special relativity have led to the idea that specific types of motion in space or specific geometries of space-time might allow time travel. A commonly discussed theory is that closed loops in space-time, also called worldlines, may allow objects to return to their own past. Another way to look at this is to use the widely accepted theory or relativity to describe the passage of time to people on earth and those traveling close to the speed of light. The people on earth will experience a much longer time period than those who are traveling do between departure and return.

Another hypothetical situation involves wormholes, which are hypothetical forms of warped space-time and are permitted by Einstein's equations of general relativity. Assuming travel was possible both directions through the wormhole, it could theoretically be possible to use the wormhole for time travel.

Concept Reinforcement:

1. Discuss the appeal of time travel.

2. List four theoretical concepts of time travel.

3. Explain worldliness and how they relate to the theory of time travel.

Chapter 45 – Extrasolar Planets

Chapter Objective:

- Apply astronomy concepts to explain extrasolar planets

An extrasolar planet, also known as an exoplanet, is a planet that is beyond our solar system. As of July 2008, 307 exoplanets have been detected and confirmed, most of which are classified as large planets that may resemble Jupiter. These planets orbit 263 stars and none of them have Earthlike qualities. The first discovery of a planet orbiting a star like our sun occurred in 1995 and was the first of many new discoveries. One of the major goals of discovering extrasolar planets is to find one that supports life in a manner similar to Earth.

The majority of these planets were catalogued by various indirect methods, such as measurement of energy, rather than actual imaging. The first extrasolar planet was confirmed in 1988, and now scientists have found more than 300 exoplanets.

Extrasolar planets were first of interest to scientists in the 19th century (the 1800s), even though they did not have the tools required to confirm their theories. The rate of discovery has increased dramatically with improvements in technology. In fact, 61 of these extrasolar planets were detected, although not necessarily confirmed, in 2007.

The most recent extrasolar planet discovered is only three times more massive than our Earth. This discovery indicates that even the lowest mass stars can host planets.

Image courtesy of NASA

How are these extrasolar planets discovered?

First, there are a few challenges to finding these extrasolar planets. First, planets do not produce their own light except when they are young. Second, they are a LONG way away. Third, they are hidden in the light of their parent stars, as Earth would be hidden in the sun's light.

There are several techniques scientists use to detect extrasolar planets. A few if them include astrometry, the Doppler method, the transit method, and gravitational microlensing,

The Doppler method involves measuring the wobble of stars (Doppler shift) due to gravitational pull from the orbiting planets.

Astrometry is similar to the Doppler method because it is a measure of the displacement of the stars in the sky resulting from the gravitational effects of the orbiting planet.

The transit method is a way of detecting periodic dips in brightness of a star due to a planet moving in front of the star during its orbit.

A technique known as gravitational microlensing came from Einstein's General Theory of Relativity and relies upon observations of stars that brighten when an object such as another star passes directly in front of them. The gravity of the passing star acts like a giant magnifying glass lens. Thus, if a planet is orbiting the passing star, its presence is revealed because the background star appears brighter. This does not work to find objects that do not emit any light.

Image courtesy of NASA

NASA has many planet-finding missions, including missions to find nearby Earth-like planets with life, terrestrial planets, and surveys of dust discs and giant planets. The Hubble Space Telescope has been invaluable in the hunt for new planets. Because it is not inhibited by the distortion caused by Earth's atmosphere, it is able to detect objects in the universe that are at a much greater distance from Earth than terrestrial telescopes are able to do.

The Spitzer Space Telescope was launched in 2003 to gather information on exoplanets and protoplanetary disks, which are where planets are formed. The infrared technology of this telescope allows astronomers to see into space in ways that optical telescopes do not allow. The Kepler Mission, which includes a satellite that is due to launch in 2009, will use a photometer (a specialized telescope that detects light changes) to measure the small changes in brightness caused by planet transits, which occur when planets move across (or transit) their stars. There are many more missions in development that will have, as one of their goals, the discovery of new exoplanets.

Concept Reinforcement:

1. Explain the concept of an extrasolar planet.

2. List and describe three techniques used to discover extrasolar planets.

3. Discuss the role of the Hubble Space Telescope, the Spitzer Space Telescope and the Kepler Mission in the hunt for extrasolar planets.

www.ingramcontent.com/pod-product-compliance
Lightning Source LLC
Chambersburg PA
CBHW062328220526
45469CB00008B/2629